D1192376

Maps of the Heavens

Maps
of the
Heavens

BY GEORGE SERGEANT SNYDER

ABBEVILLE PRESS · PUBLISHERS · NEW YORK

Front cover: Schiller's Christianized, Heaven, southern hemisphere, Andreas Cellarius (Amsterdam, 1660); Copperplate engraving, hand colored, 18½ x 22″ (Plate 54)

Frontispiece: The Astronomer (Nuremburg, 1504); Woodcut on paper, 6 x 4″; Title page of *Messahalah de scientia motus orbis*

Back cover: The sun in Leo (India, 17/18th century); Manuscript on oriental paper, 7 x 5″ (Plate 18)

Library of Congress Cataloging in Publication Data

Snyder, George S.
  Maps of the heavens.

  1. Astronomy—Charts, diagrams, etc.   I. Title.
QB65.S95 1984   523'.0022'2   84-6478
ISBN 0-89659-456-4

First edition

CONTENTS

INTRODUCTION

Since man first began to explore the world around him, the sky overhead has held as much fascination for him as have new lands and uncharted seas. The history of celestial cartography is undeniably intertwined with the history of terrestrial cartography. However, while maps of the heavens and maps of the continents were often produced by the same workshops, craftsmen, cartographers, and artists, and often appeared in the same atlases and sometimes even on the same engraved plate, the historical development of the two kinds of cartography took radically different directions.

In the history of cartography, the study of celestial maps has come to assume a footnote status, relegated to minor chapters in the classic works on maps and mapmaking. In-depth study of celestial charts and astronomical developments has become a function of science and other fields tangential to cartography. The assumption has been that maps of the world above have little to do with those of the real world below. Pieter Brueghel's painting of Icarus falling from the sky above a distinctly disinterested landscape (where dogs scratch themselves and a ship sails calmly by, as W. H. Auden has noted) is a fitting metaphor for the distinction made between maps of the two spheres.

This unusual treatment stems, in part, from some basic differences between the two genres. Whereas terrestrial maps derived from actual exploration by land and sea, charting the sky depended entirely on the limited capabilities of the naked eye. As the earth was a physical realm that could be described by pictographic symbols representing mountain ranges, forests, and rivers, the heavens were open to interpretation, to the

use of images from mythology and religion to create constellations representing the positions of the stars and planets.

In the Western tradition, constellations were part of a universal system that persisted into the early sixteenth century, and into the first printed maps of the stars by the German Renaissance artist Albrecht Dürer. Even then, the seeds were being sown for a total and irrevocable change in man's perception of the universe. Copernicus's heliocentric universe removed the earth and man from the center of God's creation, and consequently shook the foundations of religious, scientific, and social institutions.

As the Copernican revolution took hold, the heirs of the new world began to cast aside the star catalogs of the ancients. With the aid of Galileo's telescope, astronomers began to rechart the skies, discovering stars and constellations that had never before been visible. Data compiled with the help of the telescope by Tycho Brahe, Johannes Kepler, John Flamsteed, and Johann Bayer provided the basis for the development of astronomy, which in turn led the way for the advancement of all sciences in the sixteenth and seventeenth centuries. Toward the end of the seventeenth century, celestial cartography reached the zenith of artistic and scientific achievement with the appearance of Andreas Cellarius's *Atlas Coelestis*, often called the most beautiful such atlas ever produced.

In the eighteenth century, the enlightened and rational efforts of men of science combined with those of explorers to fill in the empty spaces on maps of land and sky. By the end of the century the pioneers and settlers of America and Australia were finding symbols of new world discoveries in the heavens: images of the Indian, the flying fish, the toucan, and the bird of paradise.

And now, as man's dream of physically exploring the universe has evolved to science fiction and then reality, the history of celestial cartography, in its traditional sense, has drawn to a close.

If celestial mapmaking has uniquely concerned matters of science and religion, observation and faith, it also reflects man's efforts to produce a coherent model of the universe and to establish a calendar. From the biblical story of creation to an Oriental universal diagram connecting the spirit world with the world of man and the sky overhead, from the discoveries of Babylonian astronomers to Indian interpretations, from sky-riding gods in their chariots to the twelve apostles ruling the horoscope, images symbolizing human needs and fears have been imposed on the heavens throughout history by every faith and every culture.

The charts in this book have been drawn from a variety of sources and represent different artistic traditions and different mediums and techniques of celestial representation, some of which fall outside the traditional realm of celestial charts. Because studies of celestial maps are often too narrow in scope, the range of materials here has been expanded to demonstrate the pervasive and far-reaching influence exerted by the first mapmakers of the sky. It was the astronomer who determined the duration of a year, the divisions of the earth, and the divisions of the sky. It was the celestial mapmaker who supplied the information a sailor would need to find his way across the ocean in the night without a familiar horizon to guide him. It was a map of the sun's path through the sky that provided astrologers with a system of predicting the fates of nations and kings and the personality of a newborn child. Many aspects of our lives are still affected by those who studied and mapped the skies long ago.

Richly illuminated manuscripts on vellum, hand-colored engravings, strange, simple woodcuts, and elaborately drawn figures and diagrams indicate the breadth and scope of celestial cartography. The examples following come from many times and places, representing a variety of conceptions and convictions about the way the sky looks and what it means.

Through these maps and images we are able to imagine ourselves looking heavenward in the time of Ptolemy, often from a vantage point no one before the astronauts could enjoy. Even for our space explorers, the sky never looked as wondrous as it does in these maps. Never has a living soul gazed past the tail of a dragon and around the claw of a scorpion to view the north pole of our world below, but the heavenly cartographer could imagine such a journey. Maps of the stars are exercises in creative freedom, bounded only by self-imposed limits of perspective and design. For the viewer, these maps provide an opportunity to see the sky as it has never been seen, but it is difficult to imagine that after viewing these images one could ever again look at the sky at night in the same way.

In addition to their historical and theoretical significance, these works offer an opportunity to possess something beyond a mundane understanding of the relationships and patterns of earthly life. They encourage a belief that in the infinite reaches of space there might exist answers to our own finite concerns. That awareness has come to all of us who have gazed toward heaven at the right moment; these charts capture that moment for all time.

A list of the forty-eight Ptolemaic constellations, with their respective figural meanings, is supplied here for reference when studying the plates. Constellations that have changed names, were known by other names at various times, or were introduced after the Ptolemaic system are noted as they first appeared, in the explanatory text accompanying the plates.

PTOLEMAIC CHART

Northern constellations (21)		Zodiacal constellations (12)		Southern constellations (15)	
Name (modern usage)	*Translation*	*Name* (modern usage)	*Translation*	*Name* (modern usage)	*Translation*
Ursa minor	Little Bear	Aries	Ram	Cetus	Sea monster, Whale
Ursa major	Great Bear	Taurus	Bull	Orion	Orion
Draco	Dragon	Gemini	Twins	Eridanus	River
Cepheus	Cepheus	Leo	Lion	Lepus	Hare
Boötes	Ploughman	Virgo	Virgin	Canis major	Great Dog
Corona borealis	Northern Crown	Libra	Scales	Canis minor	Little Dog
Hercules	Man kneeling	Scorpio	Scorpion	Argo	Ship
Lyra	Lyre	Sagittarius	Archer	Hydra	Sea-serpent
Cygnus	Bird, Swan	Capricorn	Goat	Crater	Bowl
Cassiopeia	Cassiopeia	Aquarius	Water-bearer	Corvus	Crow
Perseus	Perseus	Pisces	Fish	Centaurus	Centaur
Auriga	Charioteer			Lupus	Wild beast
Serpentarius	Serpent-holder			Ara	Censer, Altar
Serpens	Serpent			Corona australis	Southern Crown
Sagitta	Arrow			Piscis australis	Southern Fish
Aquila	Eagle				
Delphinus	Dolphin				
Equuelus	Colt				
Pegasus	Pegasus, Horse				
Andromeda	Andromeda				
Triangulum	Triangle				

HEAVEN AND EARTH

PLATE 1
The Creation
(France, c. 1250)
Illuminated manuscript on vellum, 15¾ x 11½"
From a manuscript of Old Testament miniatures

And God made two great lights; the greater light to rule the day, and the lesser light to rule the night: he made the stars also. And God set them in the firmament of the heaven to give light upon the earth, and to rule over the day and over the night, and to divide the light from the darkness. . . .

(Genesis 1: 16–19)

Until the end of the last century, the man of science and the man of faith were not always opposed to one another, as they have so often been in our era. Priests could be scientists (and even astrologers, as shall be seen), and scientists, such as Isaac Newton, devout theists. The tribulations of Galileo notwithstanding, many of those who preceded and followed him were men of deep religious conviction who sought to integrate faith in God's creation with their observations and understanding of celestial phenomena. Though certainly the discoveries of other cultures and religions influenced the development of astronomy in Europe, the Judeo-Christian tradition was the crucible in which the structure and validity of astronomy were tried.

This thirteenth-century illumination (plate 1) of scenes from Genesis is not a celestial chart, nor is it intended to be. It represents, however, the dominant framework of Western thought from which modern astronomy and the celestial chart emerged.

The four panels of miniatures interpret the text of Genesis describing the first four days of creation. In the first panel (upper left), God the Creator holds in his raised left hand the globe of the world, and in his right hand a mass of unformed darkness. On either side

heavenly angels stand in adoration while below His feet crouch Lucifer and the other fallen angels, or devils, depicted as gruesome monsters. God creates day and night. On the second day (upper right), God creates within the orb of the world a division between heaven, depicted as clouds, and the waters below. The third panel shows the creation of dry land, the seas, and plant life; and the fourth, the creation of the sun and moon to shine in the firmament over the earth.

The artist's simple but vibrant image of an orb containing the plane of earth surmounted by the dome of heaven symbolizes not only the relationship of heaven and earth but also the relationship of man and God. The artist provides us not with a guide to the stars but with a guide to faith. With the recovery of the astronomical works of the ancients, lost to the West during the Middle Ages but preserved by Islam, the Renaissance astronomers would begin the task of reshaping this image of the cosmos. But man does not easily abandon the images he holds in his imagination, and as he struggled to depict a "true" vision of heaven and earth, he would struggle against the limits of his knowledge and the power of his faith.

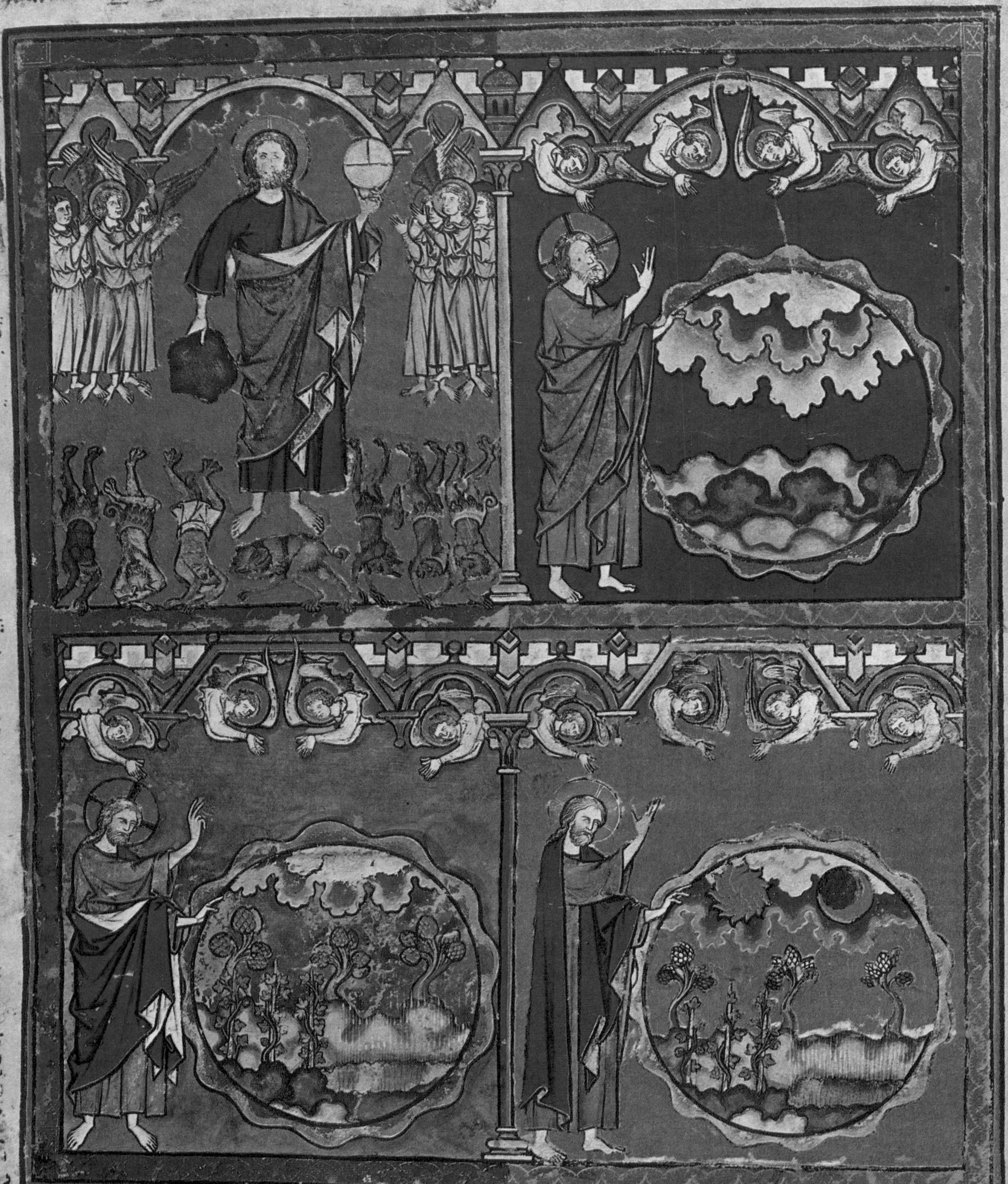

PLATE 2
Universal schema
(Nuremberg, 1493)
Engraving on paper, 17 x 12″
From Hartman Schiedel's *Nuremberg Chronicle*

In the *Nuremberg Chronicle*, representing in encyclopedic form the state of knowledge in the late fifteenth century, a graphically refined image presents the universe as geocentric, the system that prevailed in the medieval world and persisted in men's minds well after Copernicus proposed a heliocentric system in the sixteenth century, where man and earth are placed at the center of things. This geocentric system is the model that celestial charts would follow well into the seventeenth century.

As Plato proposed the idea of a supreme Good toward which everything in the universe is striving, medieval philosophers understood everything in the universe as having been created for man, God's highest creation and primary concern. In this drawing (plate 2) the earth ("terra") exists at the center of a series of concentric spheres representing the other three basic elements—water, air, and fire—all of which, according to Aristotle as he was interpolated by medieval philosophers, battle one another to find their ideal level. Fire and air are drawn upward and earth and water downward. In constant tumult, each element aspires to its perfect placement, hence the existence of change and corruption in the world. Had the elements achieved their goals, the universe would have become motionless and died.

Beyond these battling rings are the celestial spheres or crystalline rings to which are attached the heavenly bodies: first the moon, then the two lesser planets Mercury and Venus, the sun, and the outer three planets, Mars, Jupiter, and Saturn. Next comes the sphere of the zodiac, and then the sphere of the fixed stars, in unimaginable numbers. The penultimate sphere is Aristotle's "prime mover" *(primum mobile)*, considered the source of all heavenly movement and of the celestial music of the spheres. And the final sphere is the domain of God and the heavenly host. Here He sits enthroned at the top of the universe, surrounded by angels and others who gaze with Him downward at His creation. Long after the scientific investigations of Johannes Kepler, Copernicus, and Galileo posited quite a different order, this would be the popular image in the minds of men.

The sky was man's original clock, and celestial cartography developed as part of the attempt to understand that timepiece. In the age of the digital watch, the movements of the sun and moon through the heavens may have little significance in planning social and business engagements, but the ancients observed the heavens for centuries before a calendar was produced that reflected the cycles of change dictating the activities of life on earth.

A calendar is a record of universal relationships and therefore complex in its structure. The first calendars were lunar. They marked time by the phases of the moon. Each month began with the sighting of the new moon, and each year with the spring equinox. The passage of the moon occurs roughly every twenty-nine days (plus twelve hours, forty-four minutes, and three seconds), and twelve such passages total 354 days, eight hours, forty-eight minutes, and thirty-six seconds. However, since there is a discrepancy between the lunar year and the solar year of 365 days, adjustments must be made in the lunar calendar to prevent an ever-widening gap between lunar and solar cycles.

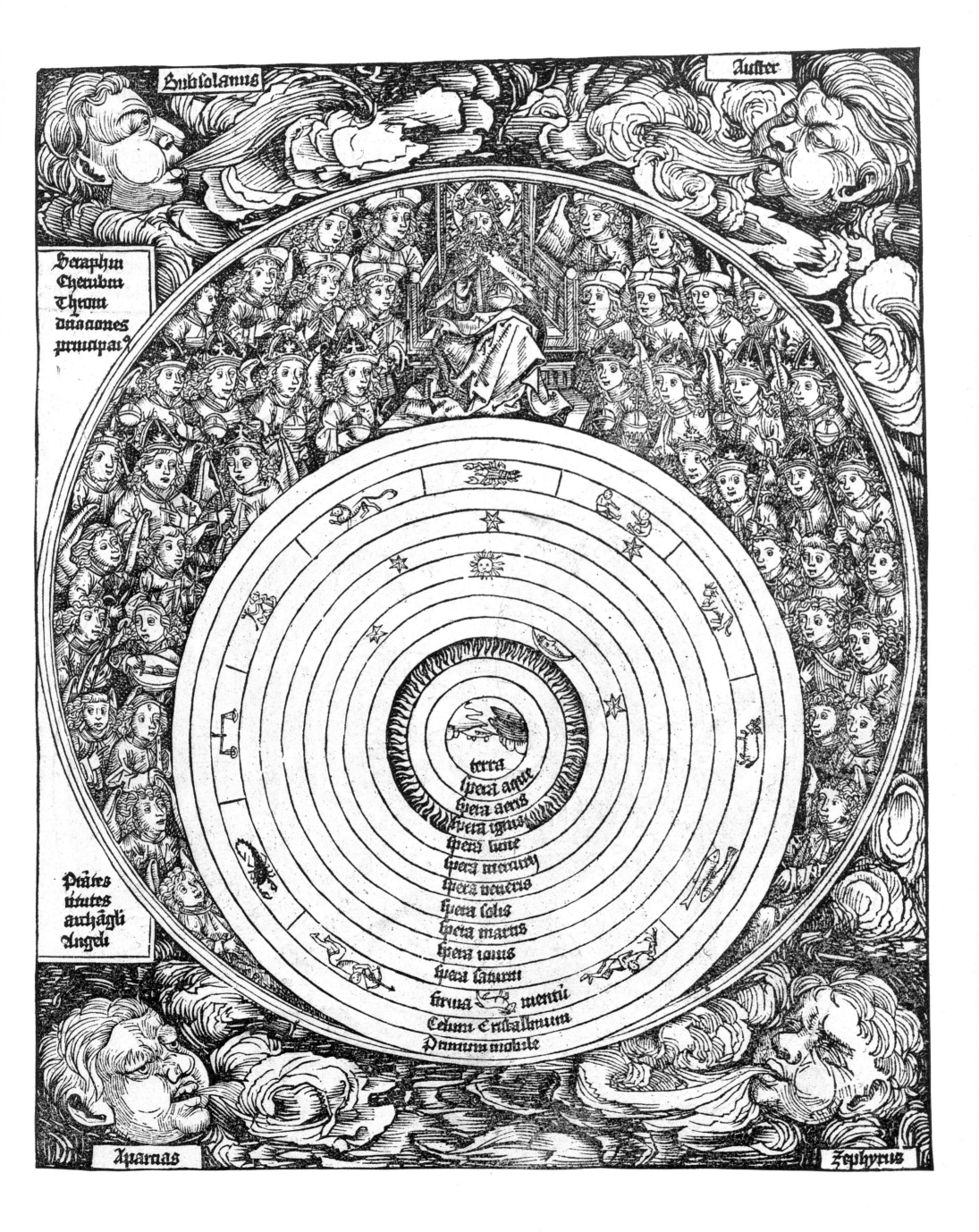

The Jewish calendar in use today is essentially lunar, with the inclusion of an extra "month," Adar Sheni, and other modifications to maintain a connection with the solar cycle. The Roman calendar was also lunar, but by the time of Julius Caesar intercalary months had been inserted so irregularly as to create nearly complete disorder. With the revisions ordered by Caesar, the lunar month was essentially dispensed with.

The calendar developed by the Babylonians in the fifth century B.C. more or less accurately combined the solar and lunar cycles to the degree that it was possible to determine the positions of heavenly bodies in the past as well as in the future. In Babylon, astrological predictions based on the sun's and the moon's positions in the zodiac developed into a highly sophisticated science. Astrology developed hand in hand with astronomy, virtually indistinguishable one from another until the Copernican Revolution made clear that astrological prognostications, predictions, and forecasts, based as they were on a geocentric system, were interpretations of celestial data derived from a system that was ultimately defunct. The belief that celestial movements influenced earthly happenings, that a relationship existed between heavenly and earthly events, however, has played a large part in the development of astrology. A significant event in the heavens—a comet or eclipse—was connected to a major event on earth, and astrological prognostications were born. Once records of celestial events were created, the temptation to associate those events with historical records made possible associations that in time came to be the basis of a system out of which forecasts and future predictions would be made.

The zodiac, the twelve constellations which mark the apparent passage of the sun across the sky, is the basic and most popularly known element of astrological forecasts, the system having been perfected in its present form by the Babylonians and inherited by Greek and Roman civilizations by the fourth century B.C. It is important to note that astrological considerations underlie many of the reproductions shown on these pages. Astrology is in many ways the motive for the development of maps of the heavens, and, that astrology persists to this day in newspapers and magazine horoscopes, points to the power and influence of this field of endeavor.

In this famous illumination (plate 3) the month of May is depicted as the time when the sun chariot moves from the constellation Taurus to the constellation Gemini. In the scene of merriment the artist reveals his familiarity with the astrological implications of the month. Venus, who rules Taurus and is the matchmaker of young couples, watches from overhead in a glorious, royal blue heaven as they consummate their relationships under Mercury, ruler of Gemini and the patron of communication. Here, a young damsel rides forth on her elegant white steed accompanied by her ladies in waiting and a handsome suitor. The artist reveals the maiden's coy, bemused expression and the ardent gaze of the gentleman as they ride through a springtime glade near the town of Riom. The others in their company, as they move along to the musical accompaniment of the heralds with their instruments, echo the royal couple in sidelong gazes and graceful gestures, even down to the playful gamboling of the little dogs following at the horses' hooves. This illustration reveals not only the relationship between time's passage and the movement of celestial spheres but also the concomitant effect of that movement on earthly life.

PLATE 3
The month of May
(France, c. 1416)
Illuminated manuscript on vellum, 11½ x 8¼"
From *Les tres riches heures du Duc de Berry*

16

Calendar making was complicated by the necessity to determine the dates of religious ceremonies and observances. The winter solstice, the spring equinox, the appearance of the full moon, and the longest day of the year were events that had to be incorporated within a calendar that also had to predict solar positions and lunar phases accurately and consistently. Calculating the date of Easter or Passover, for instance, cannot be done easily without a calendar.

Julius Caesar's Roman calendar, which essentially dispensed with coordinating lunar months, was maintained until 1752 in England. Pope Gregory XIII devised the Gregorian calendar we use today. It was adopted in Catholic countries in 1582. Its system of leap years keeps the twelve-month system in line, though our current calendar still has a mean error of twenty-six seconds per year or one day every 3,300 years.

In England, where the Gregorian calendar had been rejected out of anti-Catholic bias, the conversion to the new system in 1752 created riots in the streets. Understandably, the people viewed the government's move to conform to other European nations as a threat to their most precious possession: time. The difference between the two systems was eleven days, and the protesters, realizing this, demanded their eleven days back. They were ultimately reconciled to the adjustment, however, realizing that no time had actually been lost, only that a new way of keeping track of the days and weeks had been imposed. Yet, like the person whose birthday falls on the last day of February in a leap year, the difficulties of dealing with a calendar shift can be imagined. How strange it must have seemed to an Englishman in 1752 to face the prospect of finishing a day's labor and awake to a new morning eleven days later; how disconcerting to calculate and realize

that one's birthday would next fall more than a year from the last.

This woodcut (plate 4) from a French translation of early Greek and Latin astronomical and astrological texts, produced in 1534 by order of King Phillip, indicates the yearly cycle of earthly activities as they relate to the twelve phases of the zodiac ranged above the couple at the center of the circle. Moving left to right, from plowing the fields, harvesting and reaping under the signs of Gemini the Twins and Leo the Lion, round to the little figure warming her hands at the fire in the winter months under Pisces and Aries, the events of earthly existence conform to the celestial elements overhead at the given times. The simplest of calendars, the woodcut presents the relationship of constellation positions to the labors and duties on earth; a clear, if somewhat crude depiction, accurate enough to show a connection between the heavenly and earthly spheres— a series of circles divided into equal numbers of phases, twelve above and twelve below.

Conveniently, the twelve months of the year coincide with the twelve signs. The number twelve also has possibilities in its symbolic meaning—like the German mapmaker Schiller in the eighteenth century, Venerable Bede would nominate in the eighth century twelve apostles to stand duty for the zodiacal signs.

But anyone aware of his horoscope knows that a calendar month does not coincide with a zodiac sign. The units of demarcation we use and have used in the past relate to one or several elements of the universal system, but not to all at once. The system we create to imitate the universal has its shortcomings. Like the man who only knows how to count by twos, the method works only until someone gives him three apples.

Represented here is an eternal system which pur-

ports to be a miniature model of the universe. It is a beautiful system, comforting in its harmony of elements and in sharp contrast with the harsh realities of modern life. But this charming depiction of agrarian life, with little figures chopping wood, plowing a furrow, wielding a scythe, would disappear in the same fashion as the system in which this life is depicted. The ancestors of these French peasants would, some 250 years later, overthrow their rulers and institute a new, albeit short-lived, calendar. Long before the French Revolution and the *An I* ("Year One") revolutionary calendar, however, Copernicus would instigate a revolution by positing a new universal system which would undermine the way people looked at the sky and their activities below it. Astrology, long considered a science

indistinguishable from astronomy, would be relegated to a secondary position to that of the other sciences, since it was shown to be based on a view of the world that was inaccurate. The transition, however, would be neither smooth nor ever complete. The revolution would take a long time to be felt or understood by the common man. In the attempt to create a model that accurately reflected a universal system of cycles of time and life and heavenly movements, chaos and confusion would periodically reign. The English would riot over the loss of eleven days. Mythologies and nations would rise and fall in man's attempts to both explain and achieve the harmony which the ancients said could be heard in the music of the spheres.

The development of celestial cartography was throughout its history an attempt to combine the artistic representations of the constellations with factual information about star positions and celestial movement. The cartographer was both artist and scientist: maps of the heavens were a combination of traditional figural images as well as a reflection of the observed data of star positions recorded in ancient astrological and astronomical catalogs.

Moreover, the celestial charts and the precursors of those charts shown here, as the woodcut shown in plate 4, were attempts to express a universal relationship between the sky above and the activities of mankind below. The expression of that universal relationship—whether it was the cyclical nature of time, or the bond between man's activities during the year and the yearly movements of the heavens, or a simple statement of faith—was influenced and shaped by the religious, social, cultural, and artistic traditions within which

the artist–scientist–astronomer worked.

There were two sources of inspiration for the creation of celestial maps—written and pictorial. Both sources contributed in various combinations to the development of celestial cartography, culminating in the first star charts with both figural depictions and star coordinates (positions of the stars by longitude and latitude), produced by Albrecht Dürer in 1515 (see plates 21 and 22).

The chart of the constellations shown above (plate 5) is taken from an astronomical manuscript written by Giovanni Cinico for Antonello Petruciano in Naples in 1469. The figures are the familiar Ptolemaic constellations of the Northern celestial hemisphere depicted fairly much as they had come to be shown in the centuries since Ptolemy. Without reference to actual coordinates, however, the artist has arranged the constellations with rather considerable freedom, placing the figures in aesthetic rather than accurate groupings. Ursa

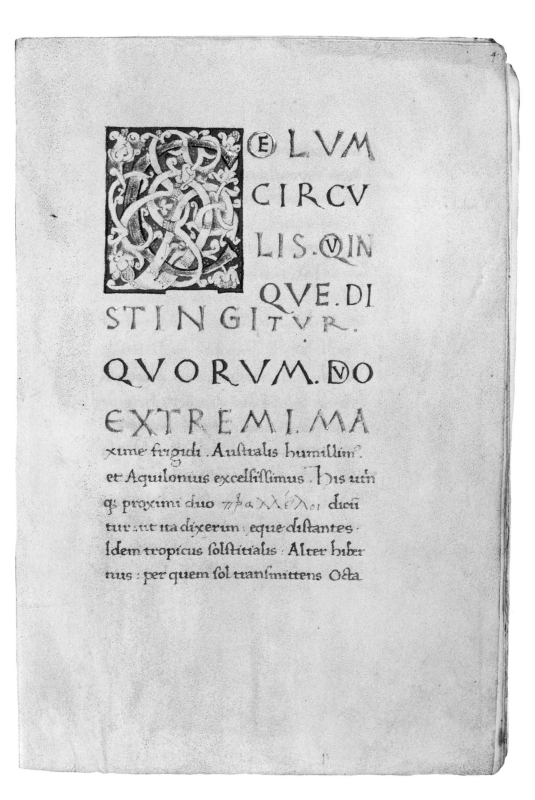

major and minor appear intertwined in Draco rather than in their actual positions, and Boötes and Virgo appear virtually head to toe to one another.

The most striking feature of this heavenly map, however, is the series of five concentric gilt rings which separate the celestial sphere. These rings correspond not to the celestial arena per se but to the regions of the earth below. In the Latin text accompanying the chart of the heavens (plate 6), the explanation of the rings is set forth. Although the rings are described as "celestial regions" *(Coelum circulis quinque distingitur)*, the regions described are in fact terrestrial. Two of these are the northern and southern poles (the northern shown in the image on page 20), which are "extremely cold" *(quorum duo extremima xume frigida)*, whereas the "southern" *(Australis)* and "equatorial" *(Aquilonius)* are excellent for habitation. The other regions, subdivided by parallels (from the Greek given in the text) form five regions in the northern hemisphere and

five in the southern hemispheres, or ten equal regions.

These "celestial" demarcations could be brought down to earth and employed to mark off terrestrial regions, and the earthly parallels indeed took their names from celestial configurations. The tropic of Capricorn and tropic of Cancer became earthly divisions created by a relationship between heaven and earth. Just as easily, then, these parallels could be circumscribed on the earth or in the heavens, as shown here: the effect was to present a universal connection between the two realms.

Text and visual images combine here to present a celestial view that reflects scientific data and pictorial images. Though this constellation depiction may seem far removed from later works, the elements of text and image are present here that would, in future creations, be combined and recombined in the crucible of the imagination, and result in the creation of celestial maps for hundreds of years.

THE SUN, THE STARS, AND THE MOON

PLATE 7
List of star positions
(England, 1490)
Manuscript on vellum, 11 x 9″
From Ptolemy's *Almagest*

More than 1,000 years before the Old Testament miniatures in plate 1 were executed, the astronomer Ptolemy (Claudius Ptolemaeus, c. 100–170 B.C.) wrote one of the earliest and most important surviving texts on astronomy. Ptolemy's *Almagest*, or *Great Mathematical* [i.e., astronomical] *Compilation*, in thirteen books (translated into Arabic as *al-majisti* and hence *almagestum* or *almagesti* in medieval Latin) is a manual covering the entire field of mathematical astronomy as it was known to the ancients.

The "Star Catalogue" of Ptolemy, contained in books seven and eight of the *Almagest*, lists 1,022 stars visible with the naked eye and gives their zodiacal longitudes and latitudes as well as their magnitudes (degrees of brightness). The manuscript presented here (plate 7) is a Latin translation of an Arabic translation of the original Greek, prepared in 1490 for (it has been conjectured) King Henry VII of England. The catalog presents Ptolemy's listing as revised in the Alfonsine Tables (prepared in the eleventh century under the direction of Alfonso X, King of Castile) and revised in Oxford in 1440 for Humphrey, Duke of Gloucester.

Three longitudes are given for each star: the first for the time of Adam (calculated as being 3,496 B.C.), the second as given by Ptolemy (here shown as being 104 B.C.), and the third for Oxford in 1440. The other columns indicate latitudes, north or south, magnitudes, and planetary *complexiones*—the planets governing the particular star. The three longitudes given are not based on improved observations but are mathematical revisions of the longitudes supplied by Ptolemy. These revisions are a result of "precession of the equinoxes."

The precession is, simply, a very gradual shifting of the plane of the earth's equator in relation to the sky, caused by the rotation of the earth's axis around the ecliptic pole in a fashion somewhat like a spinning top. (The ecliptic poles can be visualized in the same way as the earth's poles if one imagines a sphere with its equator as the ecliptic, or zodiac, surrounding another sphere—the earth—with its equator at a slight angle to the ecliptic.) This movement causes the spring equinox—the time when the sun crosses the celestial equator as it moves along the ecliptic—to coincide with a point on the ecliptic about fifty seconds of an arc earlier each year, or about one degree every seventy-one years. Since celestial longitude is the distance from the spring equinox along the zodiac, the longitude increases by roughly one degree every seventy-one years. In other words, the first thirty degrees of the ecliptic (beginning with the spring equinox), known as the sign of Aries, is today almost wholly occupied by the constellation Pisces, the sign preceding Aries.

The ancient Greek astronomer Hipparchus was the first to discover precession, but he attributed the moment to a slight rotation of the stars. Copernicus was the first to explain the phenomenon correctly. There

	longitudo	latitudo		

Que est sub concatintate pedis dextri

Que est super extremitate pedis hanis

Que est super extremitatem caude

Harum 18 stellarum in magnitudine p est una

Que sunt circa calueni et non sunt in formce

Que est ipte septentrionis a uotice capitis canis

Longior et a sunt qi essent sup liaz tra so dracis

Que est declinior ad septentrionem pedibus

Que est declinior liae ad Atzionem

Reliqua et est longior ea ad Atzionem

Sequens tanum

Sequens pucidaz q sunt sub ipst tribus

Antecedens earum

Reliqua et est decia ad andie ca q i ali tparu

Harum ii stellarum i magnitudine e sut

Stellatio ancanis et i asthere algomeista

Astenne profenni

Que est in codsario

Lucidior stellis possecio et dr pethion Procyon

et est algomersa

Harum i stellarum i magnitudine p est e t

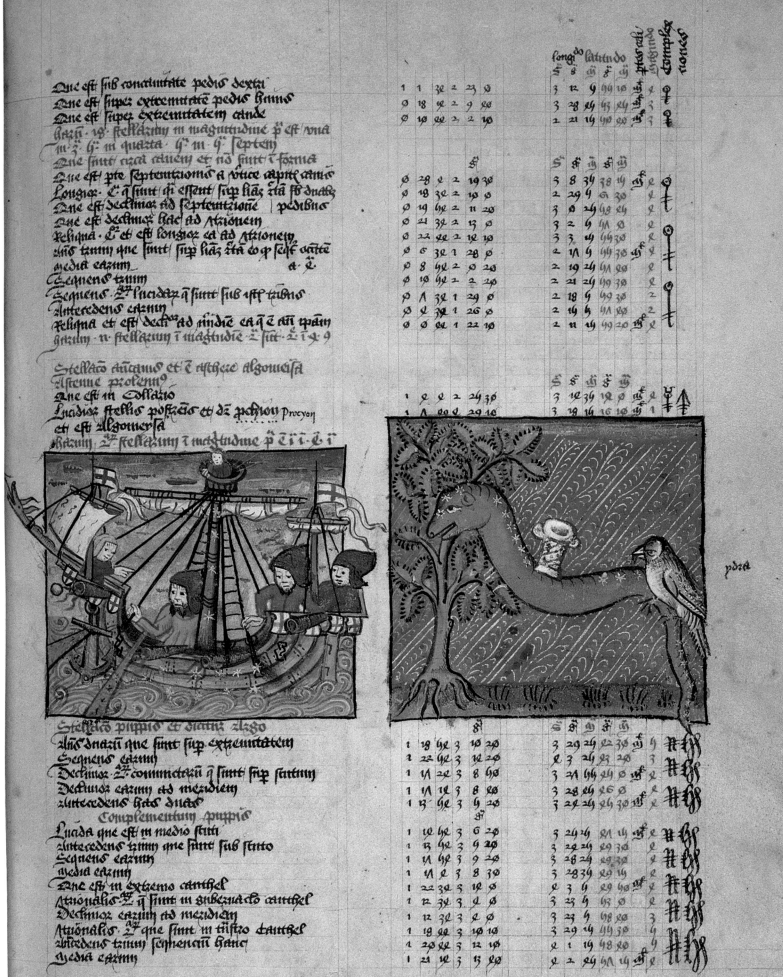

pdra

Stellatio puppio et dicitur aligo

Aliz dracru que sunt sup extremitatem

Sequens earum

Declinior 2 conuncterum q sunt sup sternum

Declinior earum ad meridiem

Antecedens has duas

Complementum puppio

Lucida que est in medio stiti

Antecedens tzium que sunt sub stito

Sequens earum

Media earum

Que est in extremo cantchel

Atzionalis q sunt in gubernaclo cantchel

Declinior earum ad meridiem

Atzionalis 2 que sunt in tistro cantchel

Antedens tzium sequentiu hanc

Media earum

was considerable disagreement for a long time, however, as to how much of a shift and therefore how much of an adjustment needed to be made. Reading star catalogs such as this one, therefore, as well as early star maps, can be a fairly complex task. To further complicate matters, Copernicus had a different system of determining longitude. The development of star maps was based on catalogs such as this one, the coordinates provided being the basis for the charts which were ultimately to be produced.

But thousands of years before the maps of Dürer or Honter, Bayer or Flamsteed, the ancient astronomers of Babylon, Egypt, and Greece were compiling records of star positions which would lead not only to an understanding of precession—an understanding that could only come after centuries of observations—but to charts of those stars in constellations which were created to describe their positions and relationships to one another.

The Ptolemaic constellations on page 25 are the southern constellations of Argo (Argo-Navis), depicted as a ship, and those of Hydra, Crater, and Corvus, a collection of proximate star groups describing, respectively, a sea-serpent, a bowl, and a crow. These configurations are pleasant as decorations, but certainly not very helpful to either the casual sky-observer or the seagoing navigator at night.

The stars have captured the imagination not only of astronomers and scientists through the ages, but poets as well. One of the earliest known "star-struck" poets was Aratus, who lived in the third century B.C. Aratus actually knew very little about astronomy, but based his poem *Phaenomena* (*Things Appearing*) on the work of the same name by the Greek astronomer Eudoxus, who had lived roughly a century before. Aratus's poem,

written for the King of Macedonia, Antigonos Gonatus, became a tremendous popular success and was translated a number of times, despite a few rather awkward inaccuracies introduced, perhaps, through poetic license or the poet's misunderstanding of the subject matter.

Aratus's work reveals the kind of fascination mankind had for the heavens, a fascination that mingled faith and superstition with what the observer saw from the earth's vantage point. Though no scientist in the modern sense, Aratus presented the information provided by the astronomers in a form easily accessible to future generations, a form which combined the data of the stargazers with the interpretations of the priesthood—expressing in conjunction with star positions the stories and myths of the constellations which had been created to express the mystical relationships of the bright points of light viewed in the night sky. Though the patterns that were seen by the ancients in the heavens may seem to us today to be capricious, the stories which accompanied them were in many cases more real than anything modern man can supply in their absence. Though later astronomers might try, no future effort could match the magic and mythic power of the heavens as reported by Aratus.

The translation here (plate 8) is by the Roman writer and lawyer Marcus Tullius Cicero (106–43 B.C.). The manuscript, produced in Italy in the ninth or tenth century, shows Cicero's translation of Aratus at the bottom of the page, with the figure of Aries above, the arrangement followed for each constellation throughout the manuscript. Within the outlines of each of the zodiacal figures are extracts from a work on the myths of the constellations by Gaius Julius Hyginus, a Latin author of the first century A.D. to whom two works on astronomy are also ascribed. The shaping of text to form a figure is a stylistic feature dating from the classical period.

Aries, represented by the ram, is the first of the

PLATE 8
Aries
(Italy, 9–10th century)
Decorated manuscript on vellum, 12 x 9″

Babyngtonf.

CICERONIS DE

e quibuf hinc fubter poruf cognofcere fultum
ima caeli mediam parum erit ut priur illae
chelę tum pectuf quod cernitur orionif
et profę confpicienf partium fub pectore clarae

ARICS

zodiacal constellations, the sign occupying the first thirty degrees of the ecliptic starting from the spring equinox, that point to where the sun, moving along the ecliptic, crosses the celestial equator. How the ram as a symbol came to be known as the first sign of the zodiac is open to much speculation, and a number of theories have been postulated. It is worth noting that the first Babylonian month, Nisan, was dedicated to sacrifice, possibly the sacrifice of a ram. In astrology, the ram of Aries is ruled by the planet Mars, its associated element is fire, and its associated part of the body, the head. In mythology, the ram was used by the young couple Phrixius and Helle to escape Hera's evil plans. While attempting to cross the sea, Helle fell off, appropriately enough, into the Hellespont, and in the end Phrixius dedicated the ram to Zeus, who placed it in the heavens. Its fleece, incidentally, was later stolen by Jason and is the subject of the immortal myth recounting Jason's adventures.

Ptolemy noted eighteen stars in the constellation Aries that were visible to the naked eye. With the aid of telescopes, Tycho Brahe increased the number of stars in the constellation to twenty-one, and Hevelius, with an improved instrument, added six more stars. Aratus's poem *Phaenomena* was, however, more than a star catalog. An artist's interpretation of not only scientific observations, but the religious implications of those observations, Aratus attempted to interpret what appeared in the sky in terms of the known and understood stories and truths about life on earth.

PLATE 9
Pegasus, Aries, and Triangulum
(Italy, c. 1450)
Illuminated manuscript on vellum, 9¼ x 6″
From Gaius Julius Hyginus's *De Sideribus Tractatus*

The iconography of the constellations evolved over the centuries on the basis of descriptions provided by Ptolemy, Aratus, and various popularizers of myths and legends. The stories of the constellations, attributed to the Roman author Hyginus, were frequently copied, translated, and illuminated with images that were to dominate the European perception of the heavens for centuries.

These exquisite miniatures are from a book devoted to the entire Ptolemaic constellation system. In detail and color they are among the finest and most perfectly

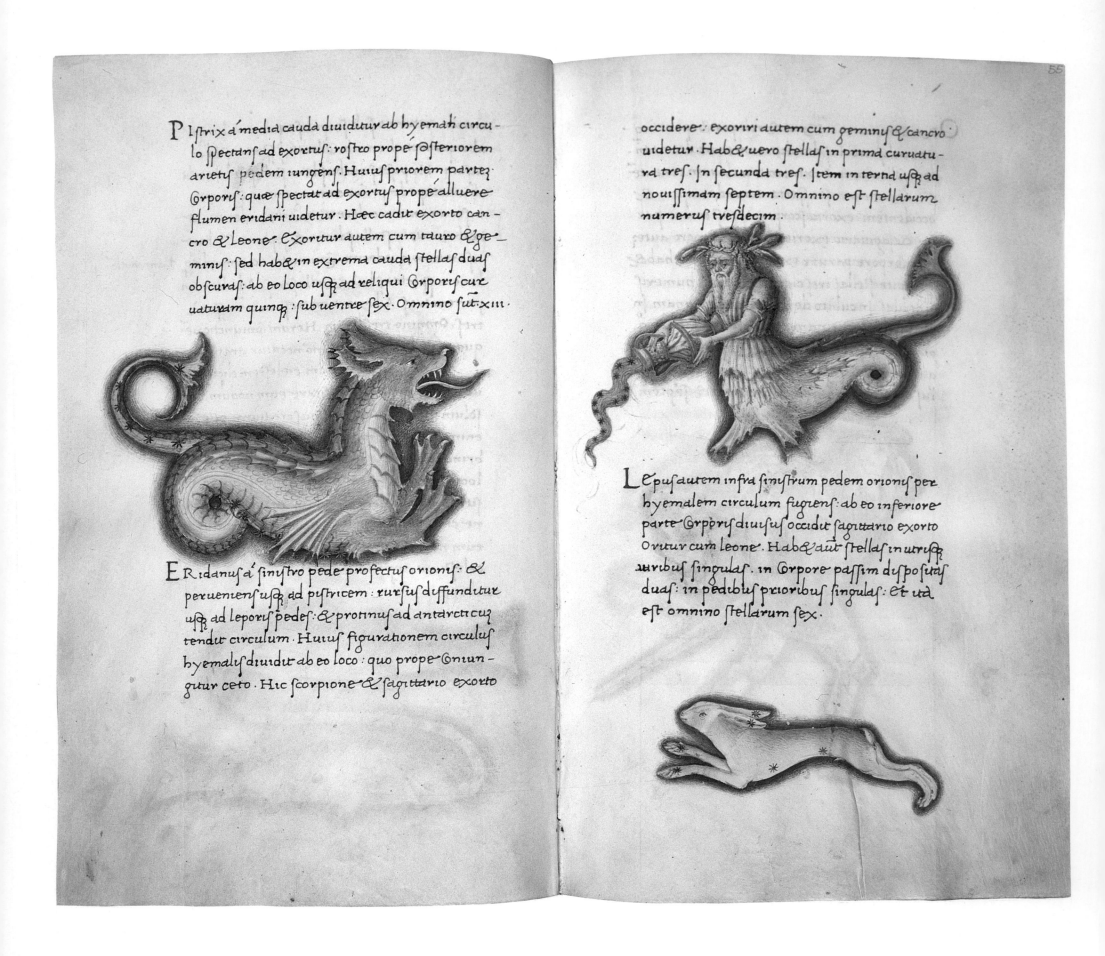

preserved astronomical miniatures in existence.

The story of Pegasus (plate 9), the marvelous winged horse that sprang from the blood of Medusa when Perseus killed her, may be found, in part, in Hesiod, the *Iliad*, and Pindar. Pegasus, who eventually found shelter in the heavenly stables of Zeus on Mount Olympus, is one of the northern Ptolemaic constellations. Aries is shown intersecting with the constellation Triangulum, also called Detoton. Though they appear near one another, Triangulum is actually found slightly above Aries in the northern sky.

Draco (plate 10), a symbol of powerful and frequently evil connotations, lies coiled at the northern celestial pole. Aquarius, the water-bearer and eleventh sign of the zodiac, is depicted as a man pouring water onto the ground, signifying the season of the winter rains. Lepus, the rabbit, appears in the southern celestial hemisphere, at the feet of Orion (plate 11), who is depicted here with Canis major. Orion, the courageous and handsome hunter, was loved by Artemis, the huntress and virgin sister of Apollo. Afraid that Artemis might fall prey to Orion's charms, Apollo sent a giant scorpion to attack the young man, who escaped by jumping into the sea. Tricking Artemis, Apollo had her kill Orion with one of her arrows while he was still barely visible in the waves, and in her grief the goddess had him placed in the sky where he was to be seen, pursued forever by the constellation Scorpion.

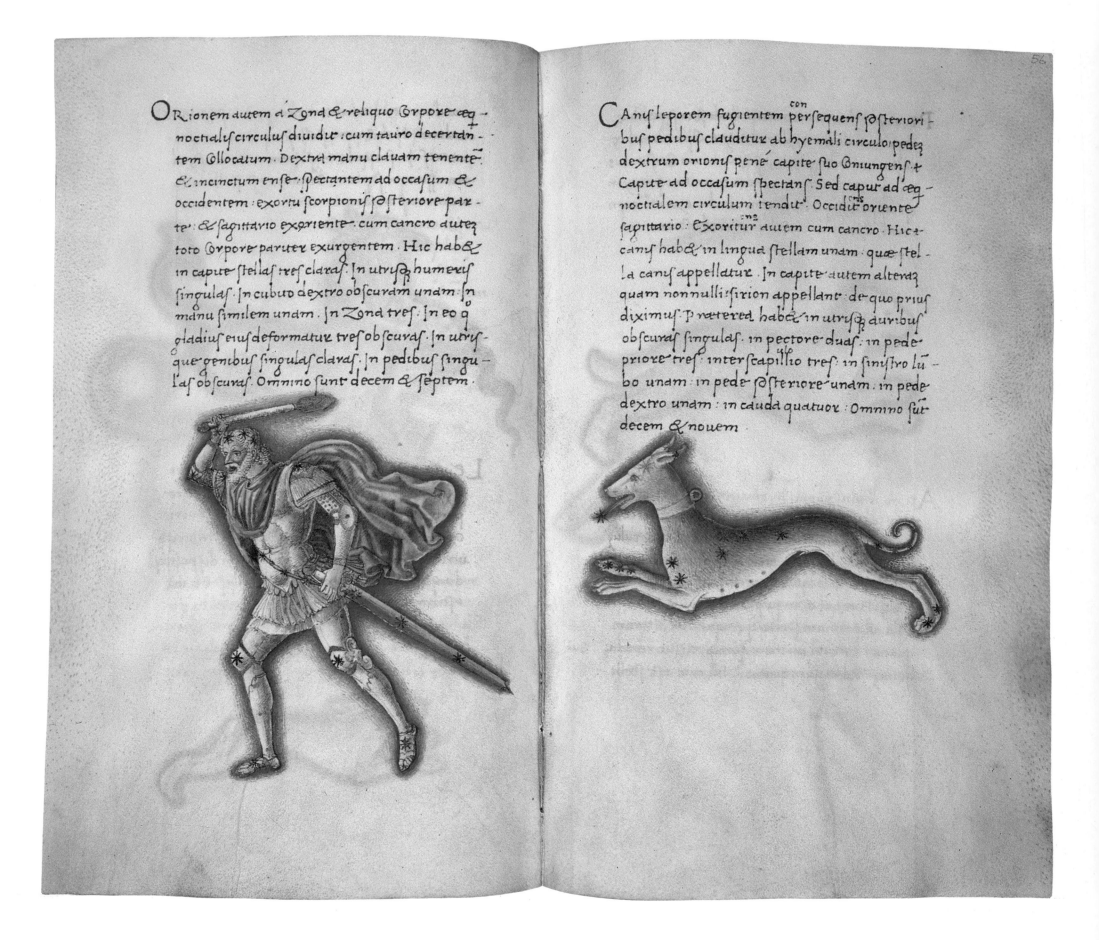

To early astronomers who charted the stars, the planets visible to the naked eye were powerful forces, rulers of the night sky. Five planets beyond the earth were known to the ancients: Mercury, Venus, Mars, Jupiter, and Saturn. Each exhibited characteristics that were recorded through the centuries. Like the sun and moon, which they seemed to serve as a sort of royal court, the planets were personified in a way that connected their positions and movements with human traits. How the planets seemed to move at great speeds, to slow down, to pass one another, or even to retrace their steps, how brightly they shown, and the various angles they made in relationship to the earth's position, convinced early astronomers and priests that they were witnessing the behavior of gods.

Images of the planets as they were known in the time of early celestial maps are represented in this Latin manuscript (plates 12, 13, 14). Commentaries on the planets are accompanied by an extensive section on the zodiac and zodiacal signs and by a large final section on the production and interpretation of horoscopes.

Venus, ruler of Libra and Taurus, more brilliant than Mercury, defines a six-month rhythmic pattern in the heavens of a graceful but vacillating character. This fluctuating celestial behavior was associated by the ancient Babylonians, Phoenicians, Assyrians, and Greeks with the flirtatious nature of the temptress who attracts and alternately shuns her suitors. She was the goddess of love, depicted as patroness of the arts, associated with youth and sensuality and beauty. In the spring, under the sign of Taurus (April 21 to May 21), she is shown wearing flowers in her hair, beckoning the coming of summer; in her other house, Libra (September 24 to October 23), she turns the leaves

and prepares the earth for winter.

Mercury, a tiny light in the sky moving in a quick little loop, became associated with the quick-witted, energetic, and cunning messenger god—Hermes, in Greek; Mercury, in Latin. The staff he carries is the caduceus, symbol of the herald and of peace. The wheels of his chariot would contain the signs of the zodiac ruled by Mercury, Virgo and Gemini.

Jupiter's movement through the heavens is majestic and evenly paced, a sure bright presence emanating from the largest of the planets. Size and brilliance cast Jupiter in the role of king of the gods, named Zeus by the Greeks and Amon by the Egyptians. Ruler of Sagittarius and Pisces, he is depicted as a king, occasionally with thunderbolts in his hand.

Mars, the red planet, has a two-year movement cycle and was seen periodically by the ancients to "flare-up" in brightness. Personified by the god of war, Mars is depicted as a soldier, a disciplined and courageous man, more mature than the lesser (and younger) planets closer to the earth, Venus and Mercury. The signs of the zodiac with which it is associated are Aries and Scorpio.

Saturn was known as the planet farthest from the earth, and as an ancient wanderer, a dim, slow moving light who passed through the sky in a thirty-year cycle. Ruler of Capricorn and Aquarius, he was the god of time, the plodding father of Jupiter who had seen all, to some the grim reaper often depicted with the scythe of death. In his movement he was the least erratic of the planets. The sign of old age, his presence in an individual's horoscope suggested long life. He was also the planet of long suffering, of tradition and forbearance, sometimes called the "Watcher of the Threshold," guardian of the outermost edge of the heavenly spheres

PLATE 12
Venus, Mercury, and the moon
(Germany, 15–16th century)
Decorated manuscript on paper, pen and ink, hand colored, 11¾ x 8″
From *Introductiones ad astrologium*

preceding the crystalline sphere of myriad stars.

The sun and moon, also depicted here, were king and queen of the heavenly bodies—orbs of divine light who in consort with one another moved in their own paths of the heavens, each influencing and interacting with the entourage of planets that composed their court. The sun, the more powerful of the two luminescent spheres, brought life to the earth. Known to the Greeks as Helius (Hyperion, according to Homer), the sun god was represented as a strong and beautiful young man with heavy curly hair and a crown of rays, driving a four-horse chariot. Rising each morning from the east and driving across the heavens in his glowing chariot he descends into the west where he is received by a golden boat which carries him while asleep back to his rising place in the east. He later came to be confused with and ultimately known as Apollo.

The moon, long associated with Apollo's sister Artemis, as well as Hecate, the moon goddess, was con-

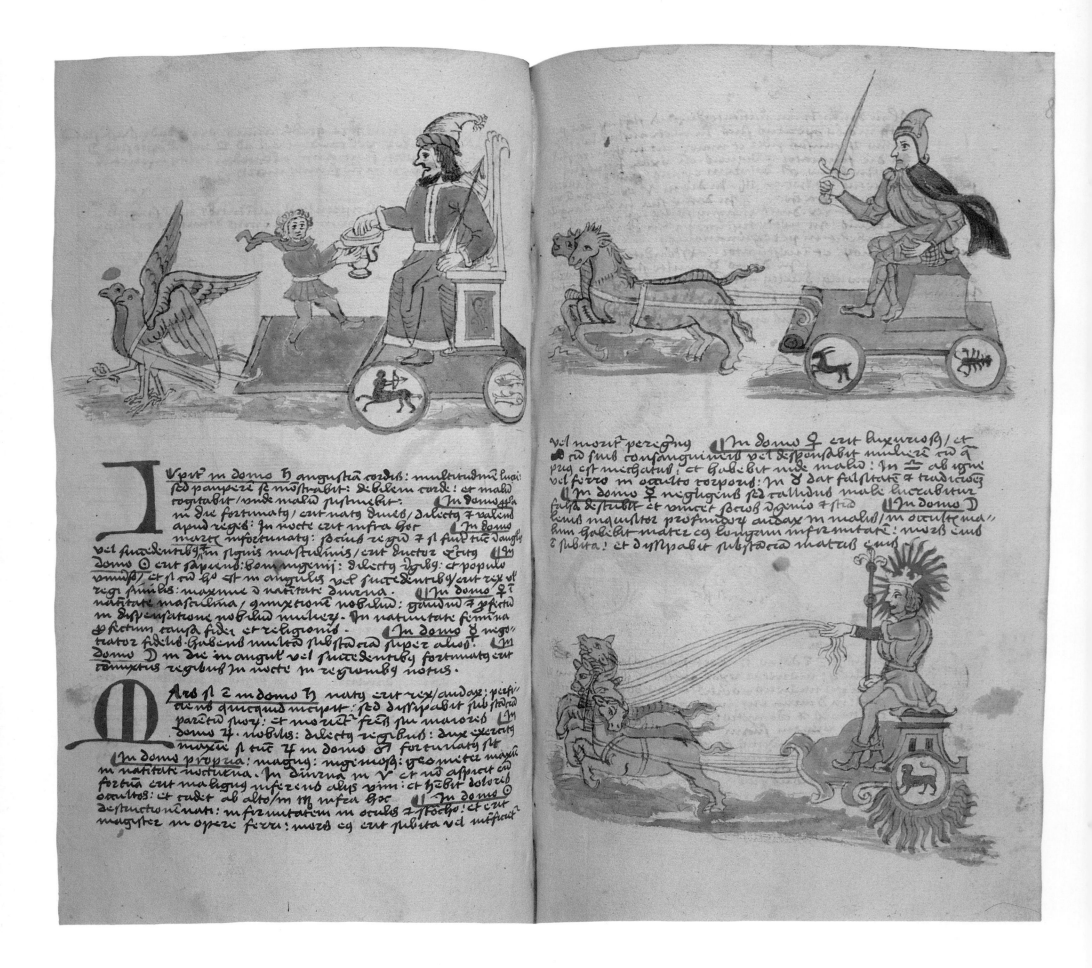

The manuscript illustrations are accompanied by handwritten Latin text in medieval script, not fully transcribable.

nected in mythology to magic and witchcraft and was believed to embody the cunning and subtle powers of women capable of driving men mad. The different phases of the moon were believed to affect life below. Both the sun and the moon, in their various stages of positions in relation to the earth, were attributed with human characteristics as well as superhuman abilities which provided them with godlike powers to rule over their subjects below.

PLATE 13
Jupiter, Mars, and the sun
(Germany, 15–16th century)
Manuscript on paper, pen and ink, hand colored, 11¾ x 8″
From *Introductiones ad astrologium*

PLATE 14
Saturn
(Germany, 15–16th century)
Manuscript on paper, pen and ink, hand colored, 11¾ x 8″
From *Introductiones ad astrologium*

fus annos gstitutos pro gradu annis, per gradus aftefland
ad radios malos vel oandam vel ad □ vel ad oppofitu ☽
aspectus: cu gplete sunt anni alboroden. natus moritur.
vel habebit perioulus simile morti

Albohali Capitulo de natitatibz quid singuli
planete signant in propeys domibz & in alijs.

Saturnus in natiuitate diurna & domus sua
fuit amicicia nobiliu & pstatoru maxi
ma, si fuit in ascudente cu parte fortune:
in nocturna labore, et morbos multitu
dinem.   In domo ♃ pulchritudinem:
in die capta substante et vitate: in nocte
cotconem nobiliu et morte patris.   In
domo martis vox lap idez: absq misericordia
abundanciam vel   In domo ☉ in die fortuna natus
et patris: in nocte dispectus vz tuisq;.   In domo ♀ de-
structione fidei natus: dilectione mulieres pauperu, et nisi
intate & impedimenta ab illis.   In domo ☿ inuesti-
gatorem scienciarum, et occulta, manifesta, vnde inuali
ndi patitur et gravabitur lingua: pessimus animo:
inuidiam patitur de eo qd non facit   In domo ☽ infir-
mitate natus & matris: et destruet substancia matris.

The Ptolemaic second century view of the universe, based on direct observation interpreted within a philosophical framework, is presented in the first two books of the *Almagest*. Though not universally accepted in the ancient world, Ptolemy's was the most generally acknowledged arrangement from Aristotle on. At the center is an unmoving spherical earth, surrounded by a sphere of fixed stars which revolves daily from east to west. Moving with this sphere are the spheres of the sun, moon, and five planets visible to the naked eye (Mercury, Venus, Mars, Jupiter, and Saturn). At the same time, in a slower motion, the moon and planets rotate on their own axes, following closely the plane of the ecliptic, the sun's apparent orbit through the heavens, so named because eclipses of the sun and moon may occur along this line.

The theories presented in Ptolemy's work, and depicted here (plate 15), required more than a single lifetime of recording the movements of the sun, moon, and planets. The literature compiled by the Babylonian priesthood covered many generations of these guardians of the temples to the gods. These priests rose to power on the basis of their record keeping, impressing the people with their increasing ability to predict phases of the moon and eclipses. Their predictions were seen by the common people as an ability to actually control heavenly events, although certain facts about the sun and moon and stars were apparent to anyone who observed the sky thousands of years before Ptolemy or the priests. Out of the Babylonian records inherited by the Greeks came such postulates as the ecliptic and the perpetration of constellation folklore passed on by the Arabs and others.

The Greeks compiled their own astronomical observations from the fifth century B.C. on, and the first serious description of the motions of heavenly bodies based on a mathematical model was the system devised

PLATE 15
Ptolemaic view of the universe
(Amsterdam, 1660)
Engraving on paper, hand colored, 18 x 22"
From *Atlas Coelestis seu Harmonia Macrocosmica*

TERRÆ
COELESTIBVS
DATÆ.

PLATE 16
The ecliptic
(Venice, 1516)
Woodcut on paper, 6 x 4″
From Cecco d'Ascoli, Francesco Stabili's *Lo illustro poeta*

by the scholar, scientist, astrologer Eudoxus (c. 400–347 B.C.) in the fourth century B.C. The anthropocentric sphere system of Eudoxus was not particularly successful, but it was adopted by Aristotle. Ptolemy's contribution to Greek astronomy as he found it was to modify the existing models of uniform circular motion and to incorporate the five planets, arranged in this diagram with Mercury and Venus between the moon and sun, and Saturn—the most remote planet visible with the naked eye—at the outermost sphere next to the zodiac.

How an astronomer charted the stars depended, to a considerable extent, on where he lived. Tycho Brahe working on the island of Hven, John Flamsteed at Greenwich, al-Sufi, Ptolemy, and others around the Mediterranean—each had a perspective of the sky that would not only influence his view of the world but would also limit, to some extent, his knowledge of the heavens. Flamsteed, for example, charted twenty of the forty-eight Ptolemaic constellations visible from Greenwich, while Brahe confined his observations to those stars he could study from his observatory at Uraniaburg.

This situation led ancient astronomers to develop a system to distinguish the part of the sky they could see from that about which they could only conjecture. These divisions would in turn be used by terrestrial mapmakers to define the earthly sphere. The efforts of the geographer and the astronomer were thus interlocked.

Recording the movement of the sun across the sky—on the ecliptic—astronomers noted the equinoxes, the two days of the year, at the beginning of spring and autumn, when the hours of day and the hours of night are equal. It was noted that on those days the sun rose due east and set due west, creating a line that was perfectly perpendicular to the earth's axis. This was also an equatorial line, and as such was adopted by mapmakers to divide the earth into northern and southern hemispheres. Also recorded were the dates of the two extremes of the sun's course along the ecliptic, the summer and winter solstices, respectively the longest and shortest days of the year. On the summer solstice, the sign of the fourth house of the zodiac Cancer (the crab) first appears, and on the winter solstice the sign of Capricorn (the goat). The corresponding parallel lines on the earth, the tropic of Cancer and the tropic of Capricorn, are the terrestrial delineations based on astronomical calculations of the sun's solstice paths.

Ancient Greek astronomers calculated that the ecliptic was at a twenty-four degree angle to the earth's equator at the solstices. Geographers therefore created parallel lines around the earth, the two tropics, above and below the equator at twenty-four degrees north and south. With these lines of demarcation, the world and the sky could be considered in their various segments, and one could theorize on those parts that were inaccessible to mapmakers and astronomers of the ancient world.

This woodcut (plate 16) shows the world in terms of meridian demarcations. Only part of the world (north of the tropic of Cancer) was charted, explored, and known to be inhabited. South of the tropic of Cancer (below the equator and extending to the tropic of Capricorn) was territory about which one could only speculate. Likewise for astronomers, the constellations of the northern sky would come first, while the southern would be filled out by later observers. In Ptolemy's work, in addition to the twelve zodiac constellations, twenty-one constellations were defined in the northern sky, and fifteen in the southern.

Ⓔ L principio che muoue queste rote
Sono intelligentie separate.

·SPHERA· ·MVNDI·

El principio che moue. In qsto capitolo tracta
dele intelligétie cioe de
angeli mouenti qsti cieli. ⁊ dicie chel principio cioe la cagióe muo

PLATE 17
The sun
(Germany, 15th century)
Manuscript on vellum, 5½ x 4"

As an object of worship and the source of life, the sun is the ultimate symbol of power and the definitive element in celestial cartography. The sun's apparent path through the sky was the key to charts of the heavens, and its placement within the cosmographical system the focus of western science from ancient times to the age of Copernicus and beyond.

Primitive man could only fear or gaze at the night sky in awe, but the sun was the largest and most influential celestial body, and therefore worthy of understanding. Long before sailors learned to steer by the stars, and centuries before the symbolism of constellations emerged, the sun captivated the imagination of man, and drove him to observe and record its behavior. Early man, marking the rising and setting of the sun, noting its positions at various times of the year and that it seemed to move more quickly at certain times, responded in terror to those moments

when the sun inexplicably disappeared, blotted out of the sky. Ultimately, after centuries of record keeping, a pattern was discerned that could be used to predict with a fair amount of regularity the position of the sun at a particular time. The path of the sun's movement through the sky, and apparently around the earth, was determined. The stars in the region of the sun's path were fixed in relationship to the sun. The solstices and equinoxes were established.

Ancient man also recorded the relationship between the sun's behavior and seasonal activities on earth, as well as the behavior of the moon and the planets. A system of interpretation developed. Human characteristics were attributed to the sun, which was seen as both human and omnipotent.

From the early medieval period onward, in astronomical and astrological works, the sun was depicted as a lordly king riding in his chariot. This iconography

mischt/ebwals die an der Sunne etwas rolant sind vnd
haben rot körnlein vnder den augen vnd habn ain klain
part vnd sind vnzarthaft vnd seind als Ogsmid vnd köch
vnd treunkhorn schier kriegsten vnd sind von kepenning
kander zc. An ayres Stund So ist ain volk gut zu
samelen der gepoen wirt der ist ain schedlig mensch der
lang get der kan nit pald schreiber komen das ee raiset die
rey waer da pey schlaft der wimbt sein schadn was vor
stoln wirt wirt wol fundn.

Ol ist der g planet vnd haist darumb sol das sy
allain scheint sur all ander sterd oder scheint allain
ob alles das auf ertrich ist ir sterd der sunen sind ve
gestalt ist seurig rot vnd kugellicht vnd ist als gro
als 8 werk als als erdrich ist vnd gibt allen sterd liecht

von orient vutz in occident vnd wirt gewont in der
hule des framamentz aber Sy scheint mitainandz dur
ch den gantzen Odiacen vnd erfult iren lauf 365 tagn
So graditt sy iren cirkel in 28 iaren die durch wandlt

became established in alchemy, magic, Tarot, and other interpretive areas as well. By the fifteenth century, when the image on page 41 was executed, the symbolism of the sun and the planets had crystallized in images that were to be used in manuscripts and printed books for hundreds of years.

The symbolism that evolved, of course, varied with different religious and cultural traditions. In India, the astronomical literature merges ancient Near Eastern, classical Greek, and Islamic elements in a distinctly Indian tradition. Plate 18 shows the personification of the sun riding the symbolic image of the sign of Leo, the fifth house of the zodiac, and the astrological sign ruled by the sun. The body of the creature is made up of intertwined spirits and writhing grotesques signifying the vitality and energy that are attributes of the sign. The figure holding the sun in his raised hand—a golden orb which itself is a symbol of

authority in many cultures—is a royal personage, a prince or raj, master of the wild beast on which he rides.

The sun is the controlling factor, as it travels through the various zodiacal signs. Its position in the sky, according to astrology, and its conjunction with the positions of the moon and other planets, determine an individual's personality, his fate, and the fate of the world at large. The sun in Leo, for instance, might—given the proper alignment with other elements—produce individuals with tremendously fiery, creative, powerful, and courageous dispositions.

This is the time of the sun's greatest power, from July 23 through August 23 according to our modern calendar, so the attributes of people born under the sign are understandable. As the primary source in the universe, the sun must also be the primary force in human existence.

The moon played a prominent role in the mythology of many ancient peoples: a feminine deity, queen consort to the sun, occupying with the sun a separate category within the heavenly system. With the advent of the telescope, astronomers turned first to the most easily observable body in the night sky. The first lunar maps were drawn by Thomas Harriot and Galileo, and the first published map of the moon's surface was produced by Matthias Hirzgarter in Frankfurt in 1643, an effort that was not entirely successful.

Johannes Hevelius (1611–1687) was the first astronomer to publish a lunar atlas. Relying on his own sightings from the observatory he had created in his native Danzig—until its tragic destruction by fire in 1679 the finest observatory in Europe—Hevelius produced by his own hand some of the most remarkable "maps" the world had seen (plate 19). The names he gave to craters and mountain ranges of the moon are still used today. Hevelius's atlas not only dispelled old notions about the nature of the moon—early concepts about its make-up, surface, the source of changing shadows and spots of darkness on its face—it also paved the way for future astronomical work that led ultimately to exploration by astronauts of the moon's surface more than 300 years later.

The son of a brewer, Hevelius managed to maintain the family business, correspond with astronomers throughout Europe, and develop astronomical instruments, besides holding public office in Danzig. His work, in many respects, reflects the sophistication of the instruments he developed, which were of great interest to his contemporaries and were published in *Machina Coelestis (Celestial Movement)* in 1673. In his Danzig observatory, the largest telescope measured twelve feet in length and had a magnification power of approximately 50X. His observations of comets and his naming of eleven new constellations added to his fame.

Hevelius's catalog of stars, *Prodromus Astromicae (Introduction to Astronomy)* (Danzig, 1690), published by his widow after his death, contains a listing of 1,564 stars, arranged alphabetically under constellation names and by stellar magnitude within the constellations. The frontispiece shows Hevelius, Ptolemy, Tycho Brahe, and others seated around a table with Urania, the Muse of Astronomy, at the center. Though he was, in retrospect, not on a level with the other astronomers, Hevelius was regarded in the mid-seventeenth century as preeminent in the field, well placed in the circle depicted in the engraving.

As seen in the Royer map [pages 120–21], the naming of new constellations often reflected the national and political allegiances of the astronomers naming them. Ultimately, however, confusion reigned as every nation vied for representation, and many constellations failed to survive.

Two such constellations are depicted in these illustrations (plate 20) for an article in the *Acta Eruditorum* (*Science Journal*), a compilation of scientific tracts and articles, published over a period of years, on medicine, natural history, astronomy, and other subjects. The description of the constellations accompanies the plate. Hevelius had created the "Scutum Sobiescianum" in honor of Jan Sobieski III, King of Poland. The "Gladii Electorales Saxonici," created by the German astronomer Godfried Kirch, symbolizes the crossed swords of the elector of Saxony, made up of nine faint stars which Kirch identified with the letters J-O-H-A-N and G-E-O-R-G, the two names sharing the common "O" at the intersection of the two swords. Whereas the "Scutum Sobiescianum" is still recognized as a constellation, Kirch's creation did not survive.

De Motibus Celestium Movilium (*The Movement of the Heavens*) written by the astronomer John Tolhopf (fl. 1440?–1480?) and presented by him to Pope Sixtus IV in 1475–76, attempts to resolve the problems inherent in a geocentric system. For instance, if the universe revolves around a stationary earth, why is it impossible to represent the movement of heavenly bodies in simple concentric rings? The sun does not rise and set in the

PLATE 20
"Scutum Sobiescianum," "Gladii Electorales Saxonici"
Johannes Hevelius, Godfried Kirch (Venice, 1740)
Copperplate engraving on paper, 4 x 6″
From *Acta Eruditorum*

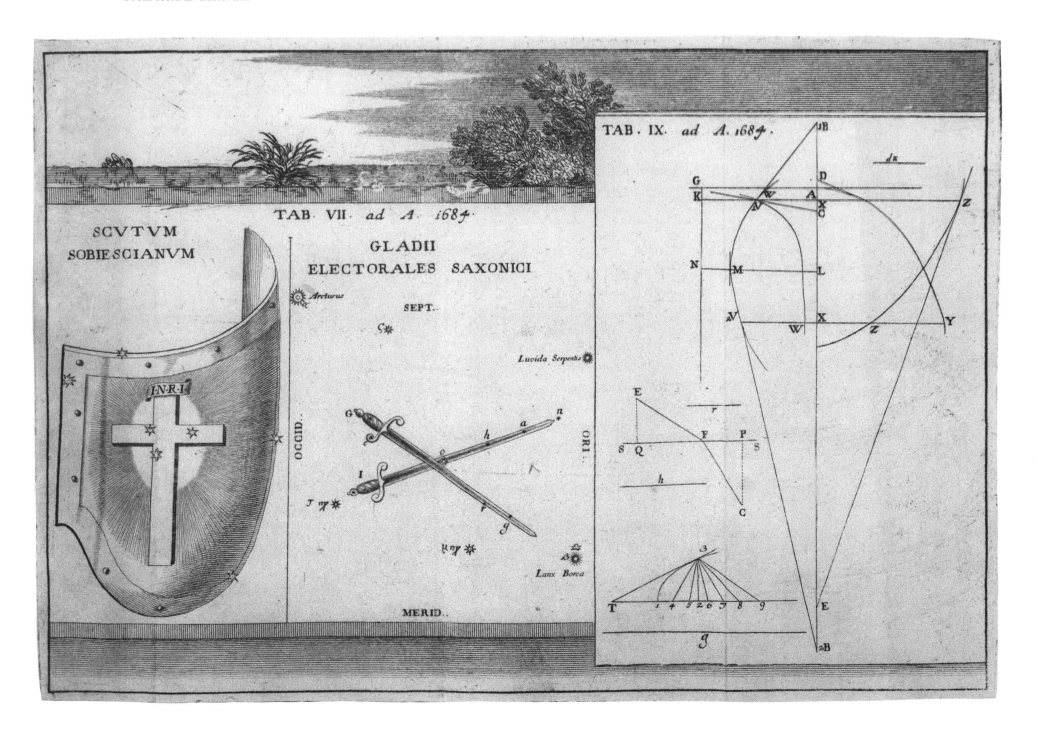

PLATE 21
Schema of the movements of the heavens
John Tolhopf (Germany, 15th century)
Manuscript on paper, 12 x 12″

same spot all year long, nor does the moon appear each night in the same position. Worse still, the major planets do not cross the sky in orderly fashion, but appear at times to stop, and even to retrace their paths, describing eccentric and elliptical routes. This peculiar behavior had to be reconciled with the design of a divine and perfect Creator, and astronomers like Tolhopf and Johann Müller (1436–1476), known as Regiomontanus, produced a variety of interesting explanations.

The task of Regiomontanus was the more practical. Called to Rome by Pope Sixtus IV to assist in revising and sorting out the calendar, he died there at roughly the same time Tolhopf's manuscript reached the Pope. Tolhopf, on the other hand, was concerned with devising a formula that would explain the seemingly erratic movements of celestial bodies. Twenty-eight different movements were depicted, contrary to some of his contemporaries who put the number at thirty-four. In the diagram that accompanied his dissertation (plate 21), Tolhopf introduced overlapping rings of movement to account for observed discrepancies. The design, though not without a certain charm, was no more or less successful in depicting heavenly movement than others before it.

Tolhopf does not, however, confine his discussion to only the celestial spheres which could be seen. He includes as well a description of the heaven of God and the heavenly hosts outside the astronomical heaven. An astronomer addressing the Pope, he was neither the first nor last man of science to exceed his sphere of knowledge for the sake of particular religious tenets. Tolhopf was aware of the necessity of placing his knowledge and theories within a context acceptable to the church. As in the case of other astronomers, it may have been that Tolhopf was struggling to reconcile his celestial theories with his own religious convictions and addressed his theories regarding celestial movement to the head of the church as the supreme authority in all matters. Ironically, as men of science like Tolhopf struggled from their presumed place at the center of God's creation to reconcile the rest of the universe to their point of view, they would ultimately develop an awareness of man's actual place in the scheme of things, and remove him as the center of creation.

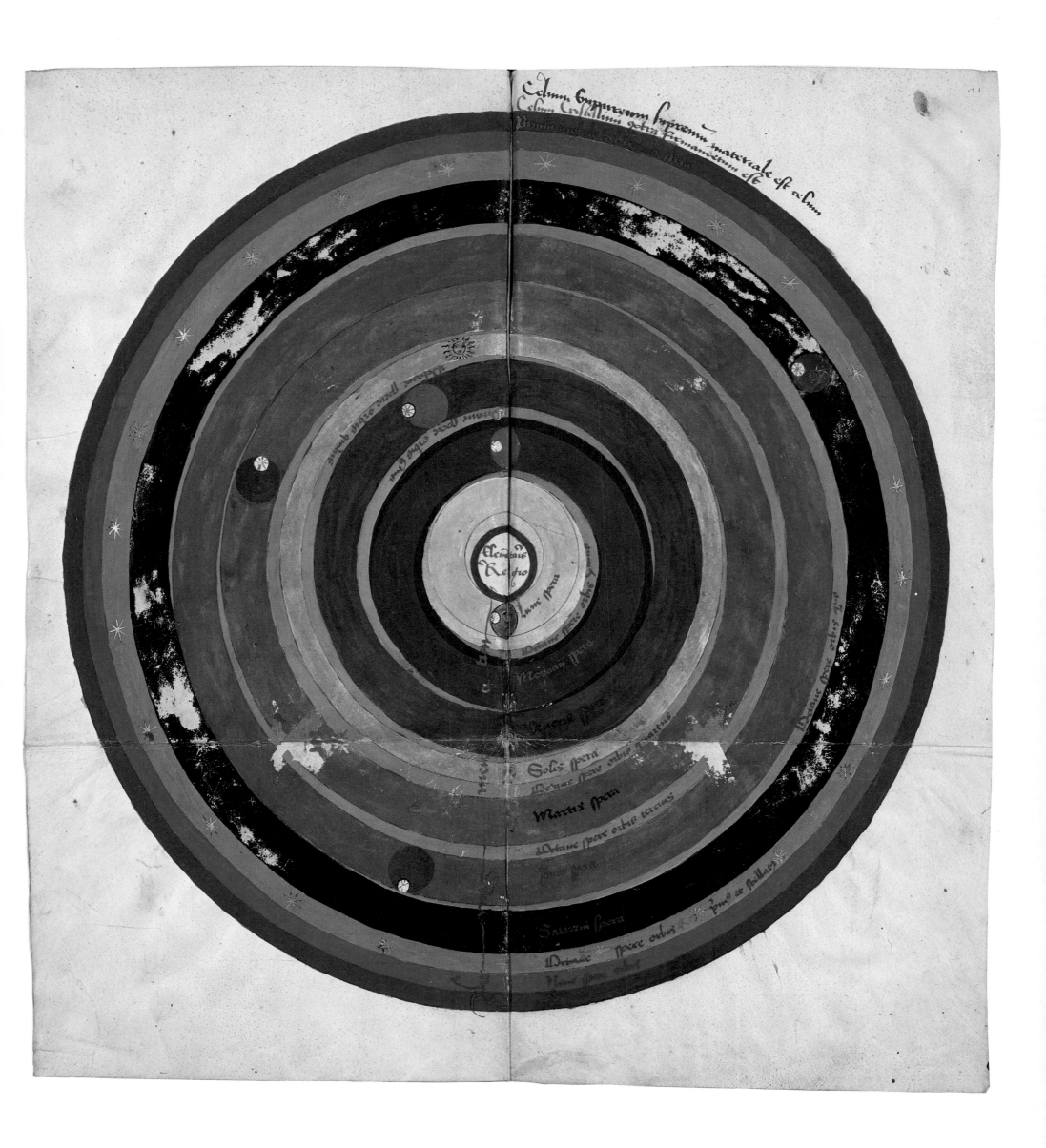

THE CONSTELLATIONS

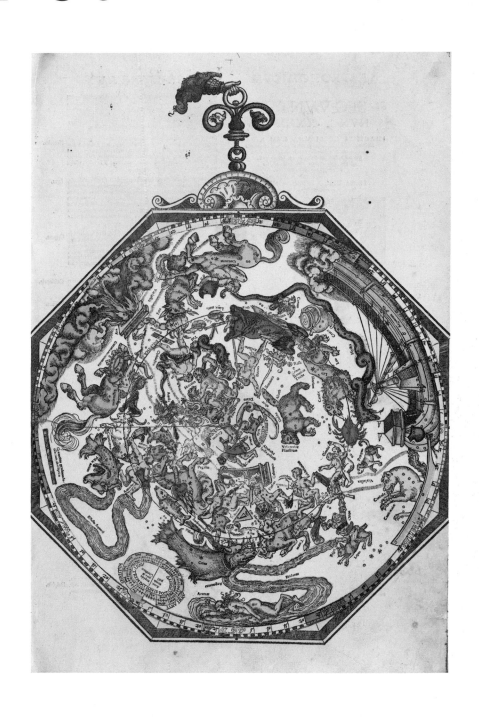

Johannes de Sacrobosco, or John of Holywood as he is sometimes referred to, was born at the end of the twelfth century in, it is believed, Yorkshire, England. He spent most of his life in Paris, where he was a professor of mathematics and where he wrote, around 1220, the work for which he is remembered, the *De sphaera (On the Heavens)*. A simple book, based on Ptolemy and the associated Arabic writings, *De sphaera* became the fundamental text on astronomy in the medieval world, required reading in all the European universities and used well into the seventeenth century. Printed for the first time in Italy in 1472, it was one of the earliest printed books on astronomy. An additional twenty-four editions were issued in the next twenty-eight years.

This illumination (plate 22) from an early manuscript version of Sacrobosco's book shows the familiar zodiac encircling a geocentric universe. Created in the medieval period, when the sky was marked out in constellations, it is an image that held currency well into the Renaissance age when actual maps of the stars were produced. Because of its widespread and long-lived popularity, *De sphaera* is a kind of bridge between the two times, the work of a thirteenth-century author that would be studied and read even up to the age of modern science. It is a reminder that transitions are rarely sudden, and that old ideas often remain popular long after new ones have taken hold.

De sphaera remained popular because it was simple and clear. It explained the meaning of a sphere, the equinoxes, and the zodiac; the basic divisions of the globe, namely the equator and the tropics; the rising and setting of the signs (the appearance of various signs of the zodiac visible at different times of the months); and the position of the earth at the center of the universe. Another equally straightforward work by Sacrobosco, written after the *De sphaera*, probably in about 1232, also enjoyed long-lived popularity. In *De anni ratione (Explanation of the Seasons)*, Sacrobosco points to the problems of the Julian calendar and presents a solution which is remarkably similar to the solution devised by Pope Gregory XIII more than 300 years later [see page 18]. Both theories conceived of a system which included leap months and dispensed with keeping track of lunar phases. In Sacrobosco, the Middle Ages foreshadowed the Renaissance and the beginnings of our own time.

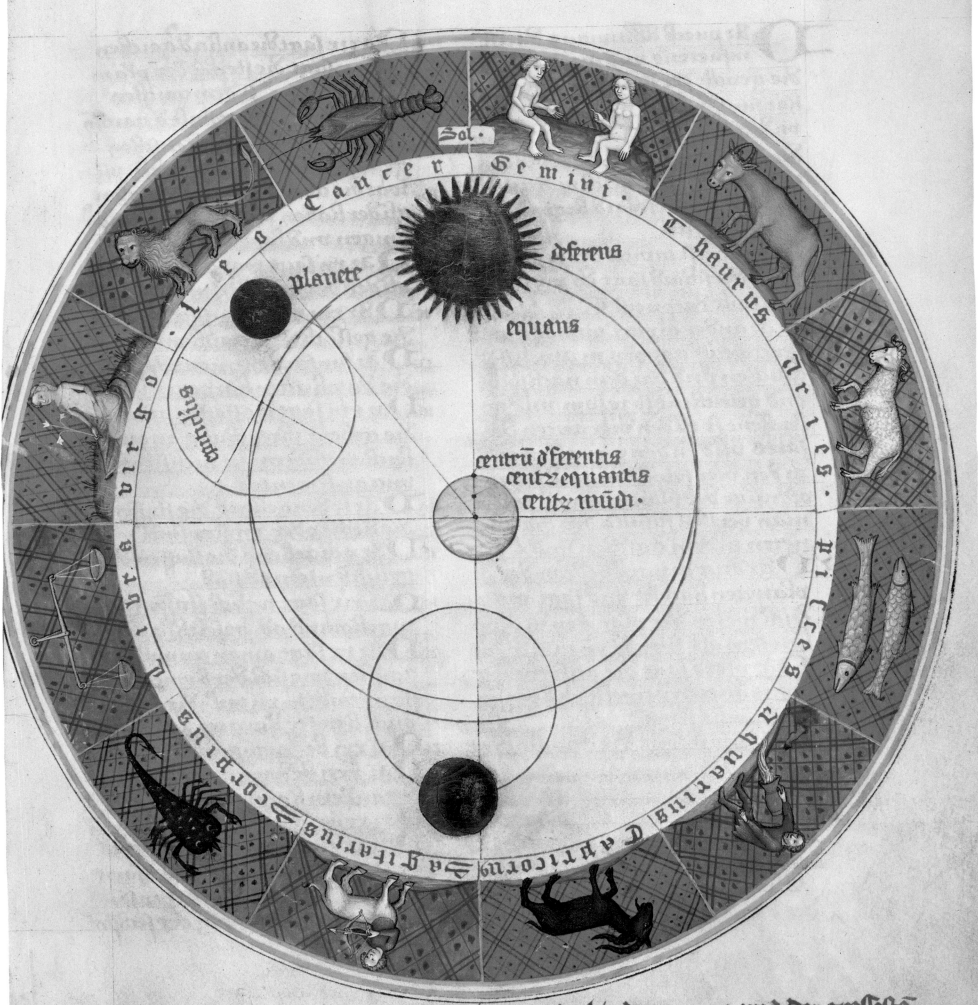

Dy figur der Epitikel der stende der richtunge der hindergeunge vnd der grossichē
hoche aller planeten vnd des mones snellichait vnd tregichait ❖

The story of the first printed maps of the stars, which provided actual coordinates for locating the stars in the heavens, is one of coincidence and irony, describing one of the great strokes of luck in the history of celestial cartography.

It is fortunate for instance, that Albrecht Dürer was, like the ideal Renaissance man, a mathematician and scientist as well as an artist, and that he came to live in the house of one of Nuremberg's greatest astronomers, after the latter's death. It is ironic, however, that although Dürer acquired the house, observatory, and scientific library in which the astronomer Regiomontanus (Johann Müller) and his pupil Bernard Walther worked, he would not be able to make use of their findings, which were not published until 1522.

Nonetheless, the Dürer maps (plates 23, 24) are the achievement of several individuals working together, and their names with their respective contributions are noted in the lower left cartouche of the southern hemisphere projection: Johan Stabius, mathematician, determined the star coordinates; Conrad Heinfogel, Nuremberg mathematician, placed the stars in their Ptolemaic positions; and Dürer drew the constellation figures.

In many respects the Dürer maps bear resemblance to celestial charts in earlier manuscripts, and sty-

PLATE 23
Northern celestial hemisphere
Albrecht Dürer (Nuremberg, 1515)
Woodcut on paper, 17½ x 17½"

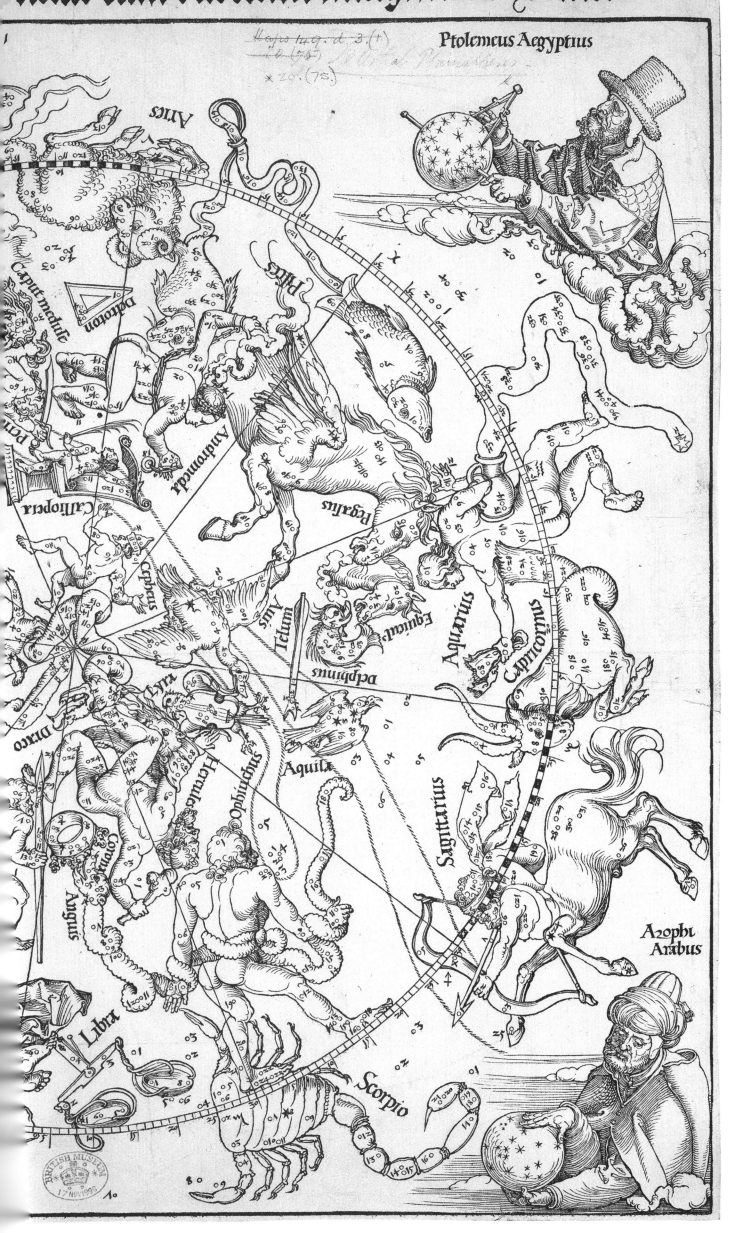

Ptolemeus Aegyptius

Aries

Caput medule

Deltoton

Pisces

Andromeda

Caliopia

Cepheus

Pegasus

Equus

Aquarius

Capricornus

Telum

Dolphinus

Aquila

Sagittarius

Ophiuchus

Hercules

Lyra

Draco

Anguis

Corona

Libra

Scorpio

Azophi Arabus

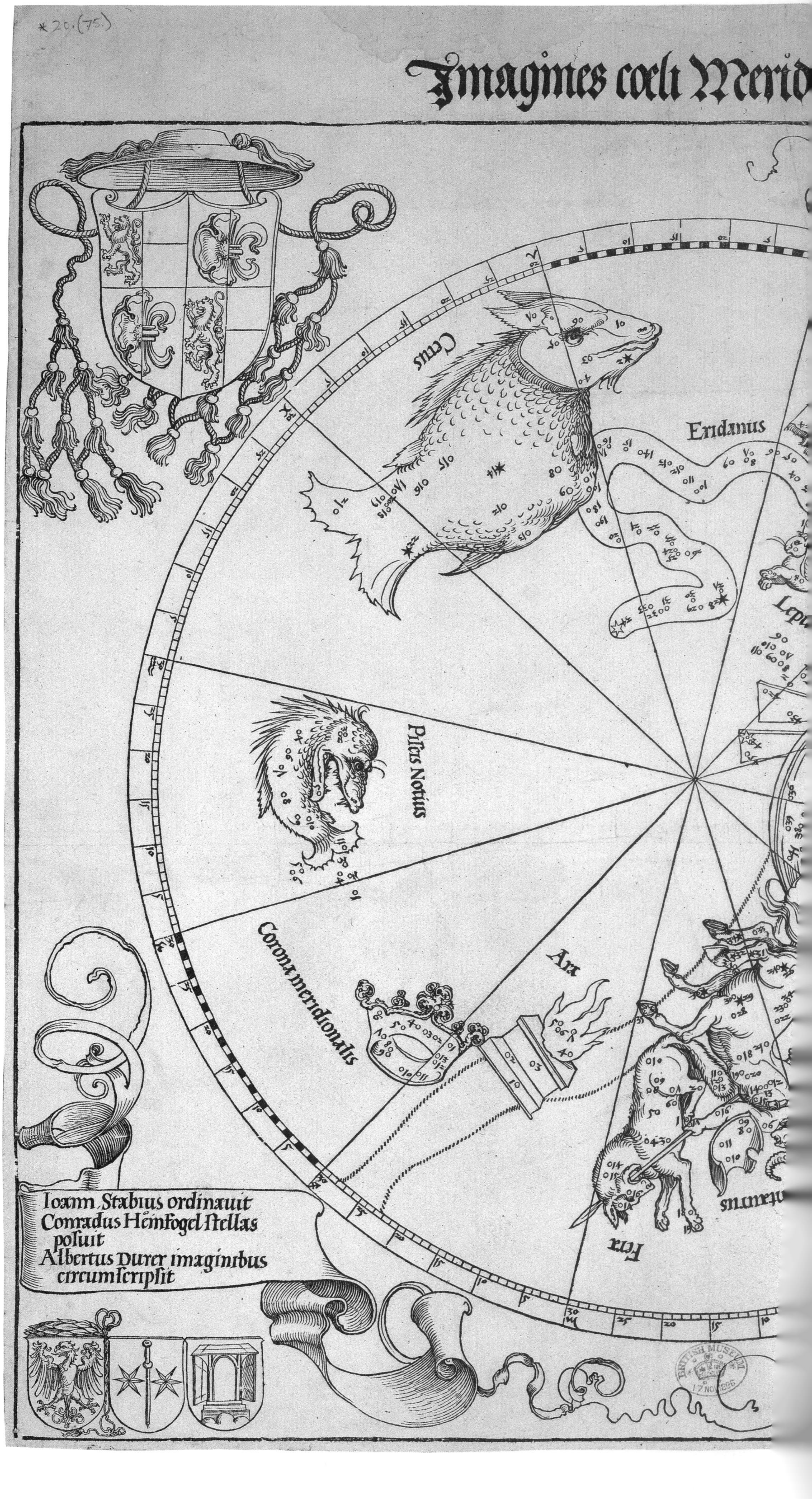

Ceus

Eridanus

Lepu

Pifcis Notius

Corona meridionalis

Ara

Fra

Ioann Stabius ordinauit
Conradus Heinfogel ftellas
pofuit
Albertus Durer imaginibus
circumfcripfit

PLATE 24
Southern celestial hemisphere
Albrecht Dürer (Nuremberg, 1515)
Woodcut on paper, 17½ x 17½"

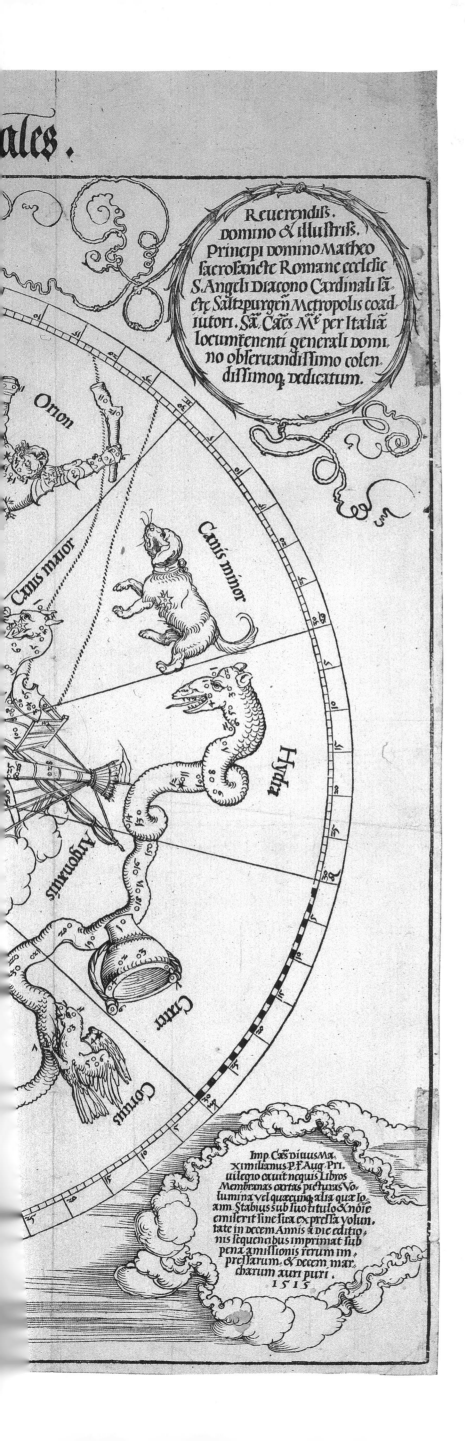

listically the designs are very much in the tradition of Ptolemaic depictions. Their significant and distinguishing characteristic, aside from their beauty, is that they could actually be used to locate and identify stars with considerable accuracy. The stars are numbered according to the Ptolemaic catalog and fixed in position with coordinates and scales. Heinfogel, in positioning the stars, would have had access to the revisions of Ptolemy found in the eleventh-century Alfonsine Tables. The catalog of Regiomontanus–Walther, with revised coordinates, was published after Heinfogel's work here, and it is not apparent that either had access to the other.

The "sources" for the maps are, in fact, indicated by Dürer's choice of astronomers to decorate the four corners of the northern hemisphere map. Ptolemy ("Ptolemeus Aegyptius") is an obvious choice. Aratus ("Aratus Cilix") is equally apt, since his poetry translated into Latin was a major source for information on astronomy and the zodiac. "Azophi Arabus" is the medieval Latin name for the Arab astronomer al-Sufi (A.D. 903–986), who was most renowned for his observations on the fixed stars. Marcus Mamlius ("M. Mamlius Romanus"), Roman author of the first compilation of treatises on astrology written in Latin hexameters, is for Dürer's time a relative newcomer to the source material on astronomy, but his work was clearly known to Dürer since the first published edition had been produced in Nuremberg in 1473–74 by Regiomontanus.

The fact that the work of Marcus Mamlius introduces little information that Dürer would not have been aware of or have known strongly, suggests that the inclusion of the author's portrait in the map was more a compliment to the publisher, Regiomontanus, than to the author's work. Here, as elsewhere in the mapping of the heavens, artists and scientists and cartographers would take the opportunity provided by their creations to further other issues. If Dürer chose to include a reference to a work published in his own city by a well-known individual, and do so out of a desire to enhance his own work with allusion to another popular local effort, then it is perhaps understandable that later other local heroes, kings, consorts, and favorites would be immortalized in the sky, through inclusion in the maps themselves and in the borders of those maps and charts.

Peter Apian, or Petrus Apianus (1495–1552), contributed most significantly to celestial cartography and astronomy through his books on astronomical instruments.

He studied astronomy and mathematics in Leipzig and Vienna, and produced a world map of considerable beauty in 1520. His first major work, the *Cosmographia seu descriptio totius orbis* (*Cosmography of the Universe*), was published in 1524 and based on Ptolemy. The work was later modified by Gemma Frisius (1508–1555), a physician and mathematician from Friesland who produced astronomical instruments and globes. Frisius was twenty-one when he published Apian's work in Antwerp in 1529, and eventually the work was translated into all major European languages. Its success gained for Apian a professorship at the University of Ingolstadt, and he was later knighted by Emperor Charles V.

Apian's second major work, the *Astronomicon Caesareum* (*Imperial Astronomy*), includes these diagrams with moving disks (plates 25, 26, 27, 28), or volvelles, as they are called. One of the more ingenious features of early printing, they are as sophisticated in their design as they are functional. In each, the hand-colored wheels are attached at their centers through the page and affixed on the verso with a silken cord ending in a seed pearl. By revolving the various concentric wheels,

movements of the planets, eclipses of the sun and moon, longitudes and latitudes, and various other data may be determined.

The first plate represents the zodiac, the second the planet Mars and its movements, the third the movements of the sun and moon, and the fourth a means of calculating eclipses. It is easy to see from these examples why Apian's book has been called one of the finest productions of the German Renaissance and, according to Professor Owen Gingerich, "the most spectacular contribution of the book-maker's art to sixteenth-century science." Artistically, the book strongly influenced the style of subsequent celestial cartography and raised standards of production. It also played an important role in the application of astronomical knowledge to astrological predictions. What led Apian to produce this extraordinary work, other than his interest in astronomical instruments, can only be conjectured, but it is clear that the results of his labors extended far beyond the realm of his interest. For the first time in the history of printing, and even today, Apian's work has achieved a status far beyond its value as a technical, scientific model book—collected and cherished as a rare and beautiful creation of the bookmaker's art, as a great work of art in and of itself, and as a source of inspiration to readers who may never have seriously studied the sky.

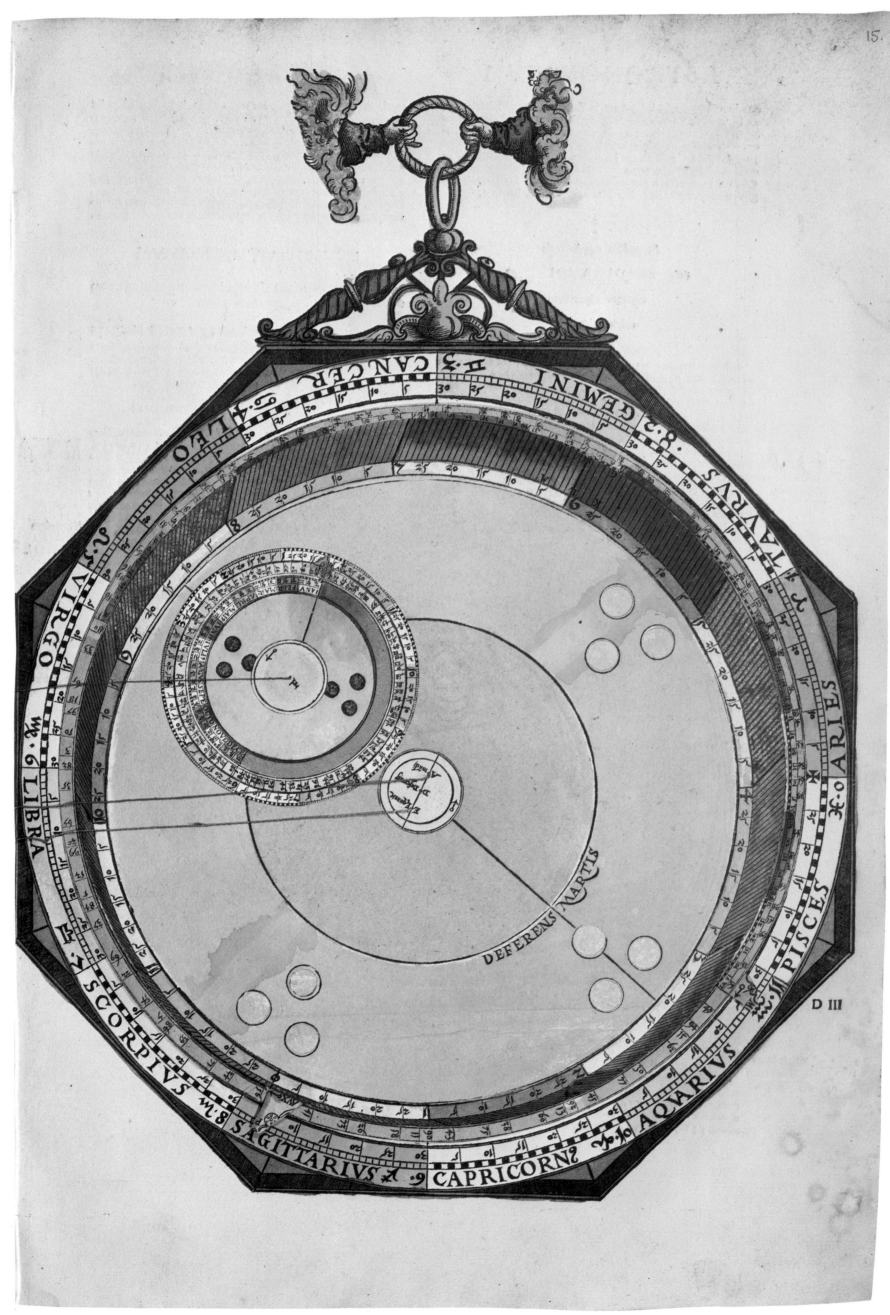

PLATE 26   Mars

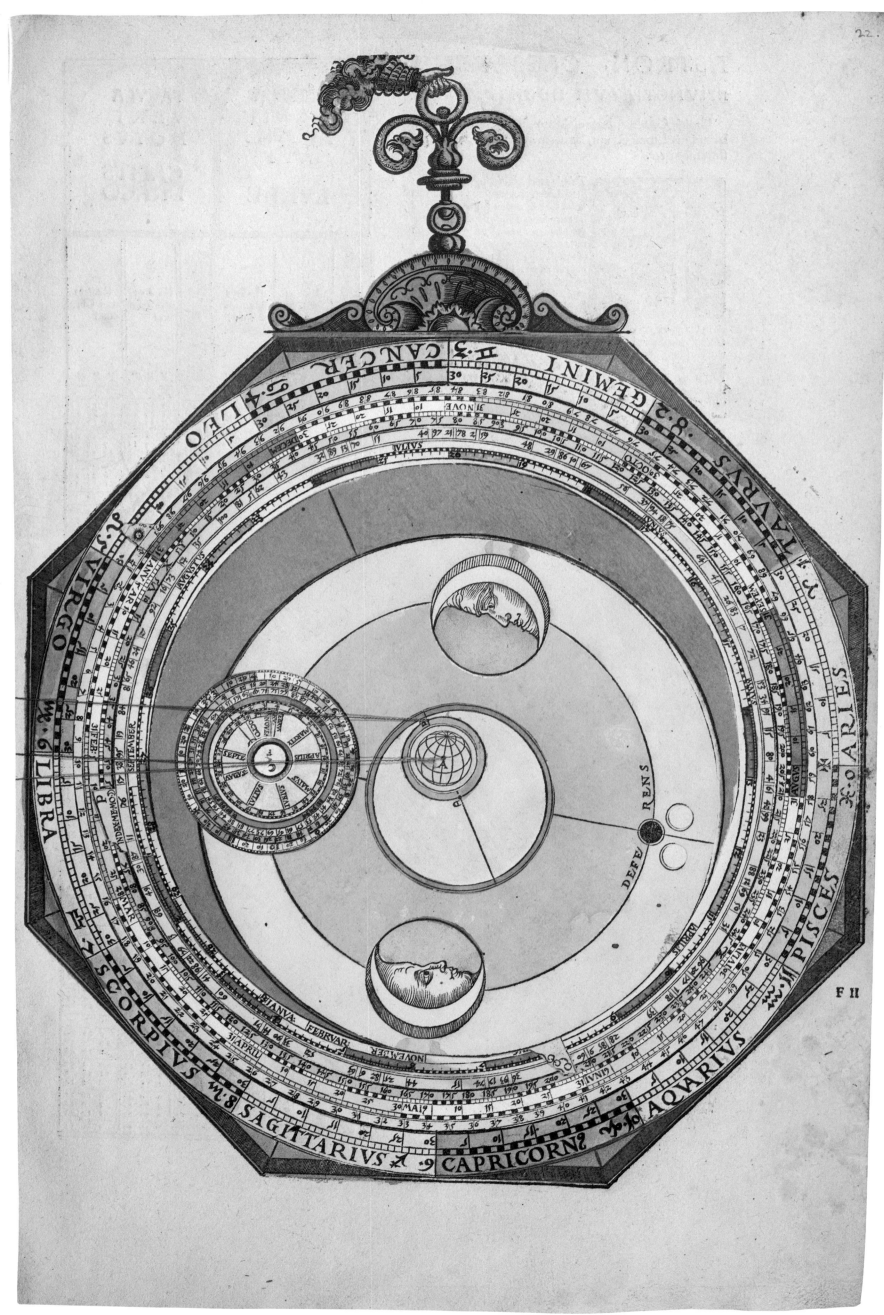

PLATE 27 The sun and moon

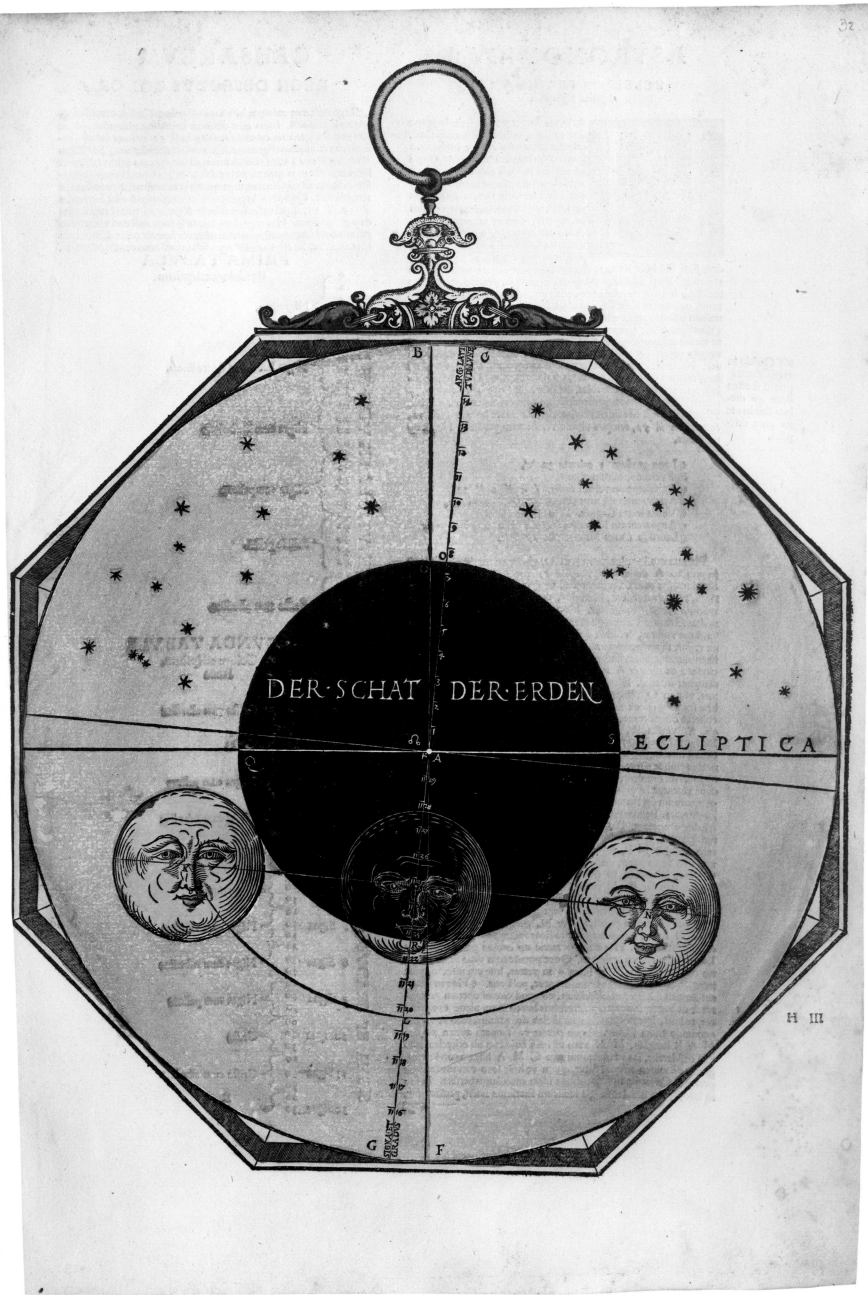

PLATE 28 Eclipses of the sun and moon

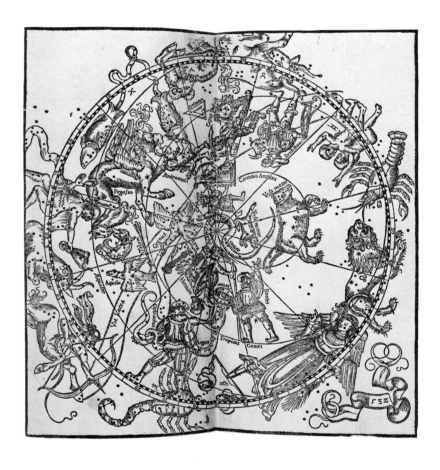

At first glance the maps of Johannes Honter (1498–1549), a German geographer from Transylvania, appear to be copies of Dürer's woodcuts of a few years earlier (plates 23, 24). In fact, the differences between them illustrate the kind of artistic freedom enjoyed by celestial mapmakers. Certainly, Dürer's maps provided the initial inspiration for Honter's work, as well as other printed and manuscript depictions of the constellations which might have been known to Honter. The improvements made by Honter, however, indicate the extent to which the artist went beyond existing models in his own creation.

Honter's constellations are depicted as they would appear when viewed from the earth, while Dürer's figures for the most part have their backs to the viewer, appearing in the sky as they would when viewed from an imaginary point outside the celestial sphere. It is obvious that such a choice of depiction would never be open to a terrestrial cartographer. In this sense, the sky is the realm of imagination, the earth of documentation. Another distinction between the two is that Dürer confined his depictions of the zodiac constellations to one plate and one hemisphere, while Honter shows the zodiac on the perimeter (the ecliptic) of each of the hemispheres, both northern (plate 29) and southern (plate 30). The advantage of duplication is that the viewer can readily envision the various skies without reference to another plate, as in the case of the Dürer maps.

Another innovation of Honter's is the use of costume for depicting various celestial figures. Honter went so far as to put his figures in contemporary dress. Dürer,

IMAGINES CONSTELLATIONVM

BOREALIVM.

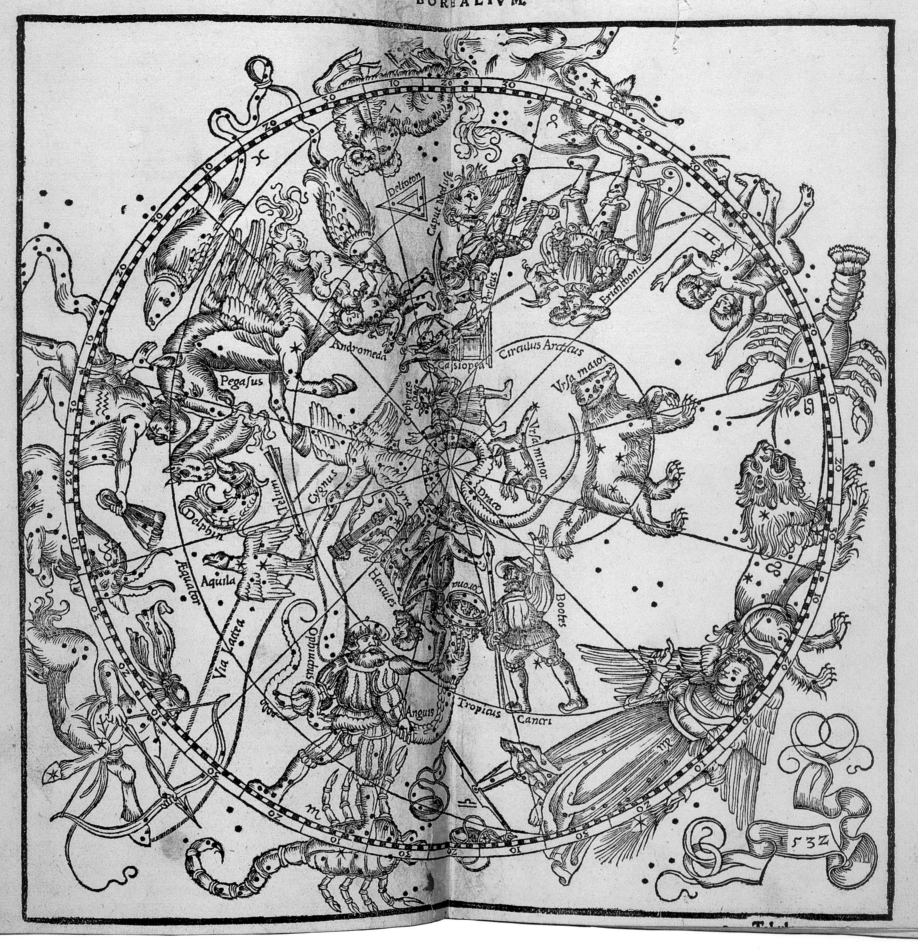

on the other hand, had attempted to conjure images of the classical world, and his figures, in various stages of undress, are in the antique manner. Honter's Perseus and Orion are soldiers, dressed in the style of the day: Erichtones (Auriga), Boötes, and Ophiucus (Serpen-tarius) are also in contemporary fashion. The figures may, in fact, seem quaint or old-fashioned to us, but one can imagine the delight of viewers some 450 years ago who might have marveled at how the heavens had become so up to date.

IMAGINES CONSTELLATIONVM
AVSTRALIVM.

PLATE 29
Northern hemisphere
Johannes Honter (Basel, 1541)
Woodcut on paper, 10½ x 10½"

PLATE 30
Southern hemisphere
Johannes Honter (Basel, 1541)
Woodcut on paper, 10½ x 10½"

THE CELESTIAL SPHERES

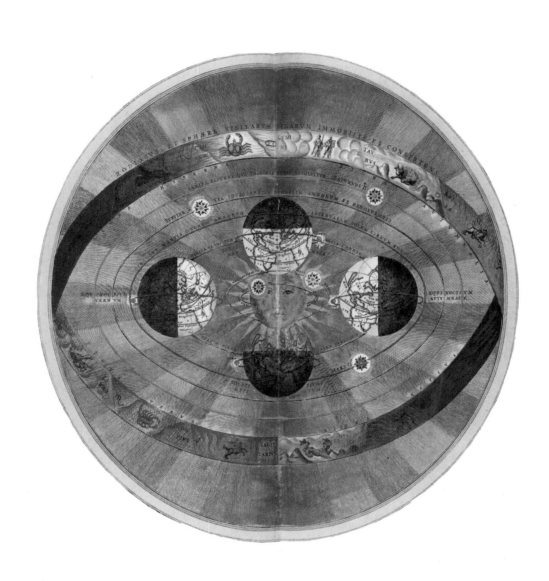

"In the midst of all dwells the Sun. For who could set this luminary in another or better place in this most glorious temple than from the place he could at one and the same time lighten the whole? . . . And so, as if seated upon a royal throne, the Sun rules the family of the planets as they circle round him. . . ." So wrote Copernicus in the first chapter of his great work, *De Revolutionibus* (*Universal Movement*), printed in Nuremberg in 1543. A simple statement on what is actually a remarkably simple concept, and yet it is difficult to think of an event or idea that had an impact comparable to that of the Copernican doctrine. Introducing the concept that the sun was the center of the universe and not the earth shattered the philosophical basis for hundreds of years of scientific, religious, and social doctrines, and though for the most part the common man might not have realized it at the time, the world suddenly and irrevocably changed.

To be sure, charts of the stars would still look the same. The ecliptic would remain the apparent path of the sun through the sky, but only apparently. Astrology, based as it was on an anthropocentric, geocentric system, would be separated from astronomy and henceforth relegated to a pseudo-scientific status. The cosmological revolution was confirmed by Galileo Galilei in his *Dialogo di Galileo Galilei Linceo matematico sopra i due Massimi Sistemi del Mondo Tolemaico e Copernicano* (*Dialogue of Ptolemy and Copernicus by Galileo Galilei*), published in Florence in 1632 (more readable, actually, than Copernicus). The engraved title page of the *Dialogo* (plate 31) shows the astronomers Ptolemy and Copernicus engaged in conversation with

the author, and above them the dedication to Ferdinand II, Grand Duke of Tuscany. The dialogue format, pitting Copernicus against Ptolemy, was a particularly effective device in defense of the Copernican system, so effective that it led to Galileo's trial before the Inquisition and his sentence to perpetual house arrest. The book—*Index Librorum Prohibitorum* (*Index of Banned Books*)—remained on the list of banned works until 1823.

Galileo's other signal achievement, and perhaps more important than any other text for our consideration here, was his *Sidereus Nuncius magna* (*On Astronomy*) (Venice, 1610), the first description of the scientific use of the telescope. In this monumental work, Galileo tells how he learned of the invention of the telescope in the Netherlands and then went about making one himself. Having produced a working model, he made observations of the moon's irregular surface, observed the constellation of the Pleiades, and described the moons of Jupiter, which he named the Medicean moons ("Medicea Sidera") out of respect to Cosimo II de' Medici, to whom the work is dedicated.

Copernicus proposed a universe in which the earth revolved about the sun, the center of creation. Galileo contributed to Copernicus's theory, and gave the world the telescope. It was the telescope that would have the most devastating effect on old ideas and theories. One could only speculate on the little that could be seen and the great that could not be seen. The telescope enabled man to exceed the limitations of his own sight, and with these achievements astronomy would lead the way toward a new definition and a new way of

DIALOGO
di
GALILEO GALILEI LINCEO
AL SER.^{mo} FERD. II. GRAN. DVCA DI
TOSCANA

Stefan. Della Bella F.

PLATE 31
Title page of Galileo Galilei's *Dialogo*
Stefano della Bella (Florence, 1632)
Engraving on paper, 8½ x 6½″

67

PLATE 32
The Copernican system
Andreas Cellarius (Amsterdam, 1660)
Copperplate engraving on paper, hand colored, 18½ x 22"
From *Atlas Coelestis seu Harmonia Macrocosmica*

thinking about the physical world and consequently about the tenets of religion. The change was, for a long time, unacceptable, and yet it could not be retracted. Galileo himself acquiesced before the Inquisition, when he was forced to deny the movement of the earth, but the truth would prevail. As popular legend had it, he muttered under his breath, "And yet it does move."

Some few years after Galileo's trial, in the more liberal atmosphere of the Netherlands, Andreas Cellarius produced this map of the Copernican system (plate 32). Within the familiar ring of the zodiac is the system that would replace the ancient one of Ptolemy. Around the planet Jupiter are the four Medicean moons, one of Galileo's contributions to the dawning of the age of the telescope.

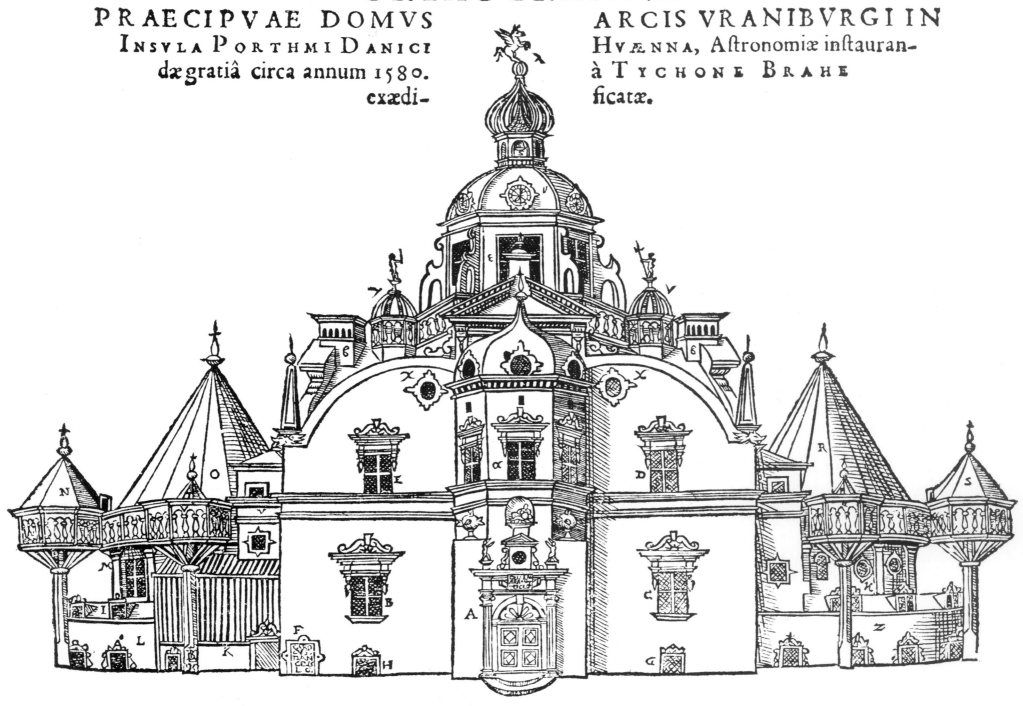

ORTHOGRAPHIA
PRAECIPVAE DOMVS   ARCIS VRANIBVRGI IN
INSVLA PORTHMI DANICI   HVÆNNA, Aftronomiæ inftauran-
dægratiâ circa annum 1580.   à TYCHONE BRAHE
cxædi-   ficatæ.

One of the most intriguing individuals in the history of celestial cartography, as much for his personal flamboyance as for his contributions to the field, was the Danish astronomer Tycho Brahe (1546–1601). An extravagant man with lavish tastes, he became a favorite of princes and lived in a fabulous castle on an island where he built his own fantastic observatory (plates 33, 34, 35). His life seems almost as incredible as the myths and legends of the constellations he devoted his life to charting.

Even as a youth making his first recorded observation of a conjunction of Jupiter and Saturn, he was aware of the lack of accurate data regarding the positions of the stars and planets. His early interest in this and astronomy as a whole led him to study everything he could find on the subject. He traveled to meet and work with astronomers across Europe. In 1572, his interest was aroused by the appearance of a new star in the heavens, in the constellation Cassiopeia, which remained visible for eighteen months. He made constant observations of the phenomenon and published his findings in his work, *De Nova Stella* (*The New Star*), in 1573. Later, Frederick II of Denmark provided him with the island of Hven, in the sound off Copenhagen and Elsinore, where he built an extravagant palace and observatory complete with gardens, printing press, workshop, chemical laboratory, library, custom-built instruments and machinery. A few years before his death, having fallen into debt and out of favor at court, he was forced to leave his observatory, which he had named Uraniaburg ("the tower of heaven"). Rudolph II, the German Emperor, then invited him to Prague where he was installed in a palace and continued his observations until his death.

Tycho Brahe's significant contribution to astronomy was in the development of systematic and refined observational data for star positions and planetary movement. In addition, he studied comets and novas (new stars), computed solar and lunar tables, and determined an accurate value for the rate of precession of the equinoxes. In all, it was a rich legacy of research that could be applied to the charting of the heavens.

Tycho Brahe refused, however, to accept the Copernican doctrine in its entirety, developing his own amalgamation of Ptolemy and Copernicus. His basic objection, for the most part on religious grounds, was to the idea that the earth moves. His system, later depicted in Cellarius and elsewhere, has all the other planets revolving around the sun, while the sun revolves about the earth. He never fully developed the theory, and his pupil Johannes Kepler concentrated his energies, after Tycho Brahe's death, on the more valuable and practical aspects of his master's work.

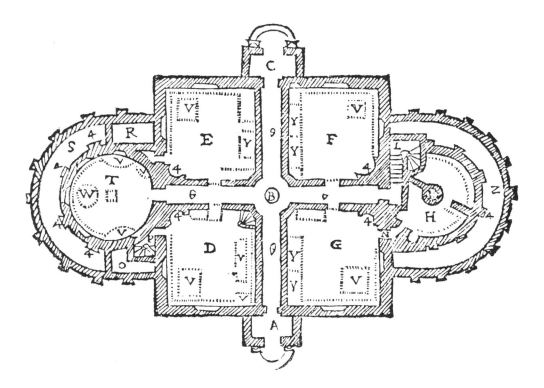

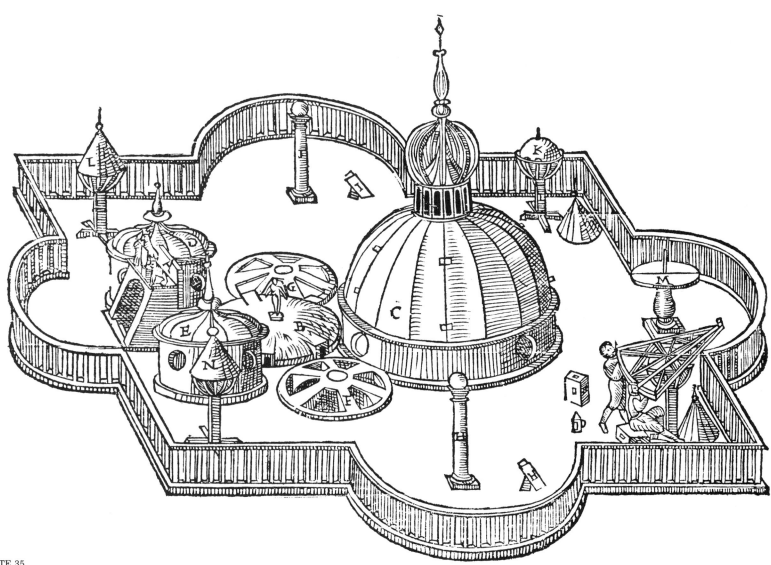

PLATE 36
Frontispiece to Johannes Kepler's *Tabulae Rudolphinae*
(Ulm, 1627)
Engraving on paper, 12¼ x 8½"

The elaborately engraved frontispiece to Kepler's *Tabulae Rudolphinae* (plate 36), published in Ulm in 1627, appears at first sight incongruous within the slim folio volume of the first edition—the busy image of Urania's temple in contrast with the following pages of narrow columns containing logarithmic tables and calculations. Yet, perhaps more successfully than the maps of the heavens which would ultimately be created from Tycho Brahe and Johannes Kepler's data, this frontispiece symbolically reveals the advancement of astronomy by the seventeenth century. Given the trials and tribulations Kepler had endured in bringing the *Rudolphine Tables* to print, it was his wish that a fine engraving be created to symbolize his labors and his triumph. His friend Wilhelm Schickard prepared a design but the final product was far more complex than Schickard's original.

The frontispiece presents the history of astronomical study symbolically in all its stages and culminates in the work for which it was designed. The temple of Urania is supported on the twelve pillars of the zodiac, of which ten are visible in the engraving. In the background the pillars are crude tree trunks, representing the earliest beginnings of astronomical observations, and by one a primitive astronomer views the heavens through the simplest tool of all, his own hand, with which he calculates angles of altitude and heavenly positions. Second are the pillars of rough hewn stone, followed by columns of brick, hung with early astronomical instruments by which the ancient astronomers Hipparchus, standing on the left, and Ptolemy, seated on the right, appear with the tools of their time. In the center, Copernicus sits next to a single classical Ionic column engaged in conversation with Tycho Brahe, dressed in a long fur-trimmed cloak, standing by an elaborate Corinthian capped column. Tycho Brahe points to the geocentric system on the ceiling of the temple and says to Copernicus, *"Quid si sic,"* or roughly translated, "How about that?"

The panels at the base of the temple depict the events which culminated in the publication of the first edition of the work. In the center appears the island of Hven, the site of Tycho Brahe's observatory. On the left Kepler is seated at his work table, a model of the temple's dome by his side. On the right printers prepare the sheets of the book at their press. Above, the dome of the temple is surmounted by the imperial eagle tossing golden coins from his beak, which shower the temple and lay about the floor below—an appropriate symbol for the munificence of governments who support the sciences, but perhaps also something of an ironic symbol since Kepler in the end financed the publication himself. More appropriate, if not a subtle symbol of Kepler's efforts, is the completion of the temple with its dome, underscored by the image of Kepler and the dome's model below, clearly indicating it was Kepler who completed the work begun by Tycho Brahe.

At the dome's edge are six goddesses who symbolize Kepler's ideas and represent his various endeavors and writings. On the far right is Magnetica, holding a lodestone and compass, symbolizing Kepler's theory of magnetic force that he believed controlled the planets. Next is Stathmica, goddess of the law of the lever and the balance which she holds; followed by Geometria; Lograithmica; a goddess with a telescope; and another who holds an orb casting a shadow, all representing various aspects of Kepler's work in astronomy, optics, and his theories of planetary movement.

Kepler's frontispiece represents the progress of astronomy, from the symbolic pillars of its primitive beginnings to the fine Corinthian column of Tycho Brahe's work. And over all the dome of Kepler completes the whole—the publication of the *Rudolphine Tables*, which for its accuracy and precision provided the observational data necessary to bring astronomy into the modern age of technology and to further science.

TABULÆ
RUDOLPHI
ASTRONO
MICÆ

HIPP
ARCH
VS.

COPER
NICUS.

TYCHO
BRAHE.

PTOLE
MÆVS.

MYSTER·COSMO
ASTR·P·OPTICA
COM·MARTIS
EPIT·AST·COP

INSVLA HVENNA DANIÆ.

Georg Celer Sculpsit Norimbergæ

Johannes Kepler was born in 1571 near Stuttgart. The sickly child of impoverished parents, he was fortunately provided with an education by the duke of Württemberg, and managed to continue his studies at the University of Tübingen. It was there, while studying theology, that Kepler learned of and was converted to the Copernican doctrine. He consequently found himself unable to continue his theological studies and took instead a position as lecturer in astronomy while he worked on and eventually published his astronomical studies. As a result of his labors, he came to correspond with and work for Tycho Brahe, ultimately succeeding him as Imperial Mathematician to Emperor Rudolph. Near the end of his life he published the *Tabulae Rudolphinae* (*Rudolphine Tables*), a star catalog of his and Tycho Brahe's observations and data.

Plagued throughout his life by ill health, money problems, and religious intolerance, Kepler pursued his astronomical inquiries with admirable diligence. He practiced astrology, producing astrological predictions and horoscopes in order to supplement his income; he is reputed to have said that Astronomy would starve if her sister Astrology did not feed her. It is reported that at one point he hurried to Württemberg to help his mother, who had been taken to trial on charges of witchcraft. In 1630, having traveled to Prague to demand back payment for services to the Emperor, he was taken ill and died, and was buried outside the city walls as a pauper. All signs of his grave vanished in the Thirty Years War.

The diagram presented here (plate 37) is from Kepler's *Mysterium Cosmographicum* (*Mystery of the Universe*), published in 1596, which is the key to Kepler's approach to astronomy and his importance in the field.

From the beginning Kepler had maintained a belief that God had created a universe which existed in perfect harmony and that the key to that harmony existed in relationships which could be discovered among the movements of the heavenly bodies. The Copernican system appealed to Kepler because of its simplicity, and he set out to show that simple arithmetic could be used to represent the relationships of the various planets within that system, in terms of their distances one from another and the sun, and in their movements. He attempted to create various models and theories which would, in light of his observations, produce a simple solution, but none seemed to work.

In this diagram, Kepler attempts to show that the relationships of distance amongst the planets can be represented in terms of five different regular solid forms (cube, globe, pyramid, tetrahedron, and decahedron) inscribed and circumscribed within one another, and that the radii of these various solids are in fact roughly proportional to the distances of the six planets from the sun—at least proportional to the distances which Kepler had been able to deduce from his and other observations of the time.

Kepler saw in this proportional relationship a key to the meaning of the universe, and the model was, as he had hoped, essentially simple and perfect. Of course, the proportions determined were *not* perfect, but Kepler blamed the variations on faulty or inaccurate observations, and set out to perfect his observational data. His search for accuracy led him to Tycho Brahe, and the results of his labors led to a place in the field of astronomy far greater than the value of the model he set out to perfect.

Kepler's data and research, his work on planetary motion, and the *Rudolphine Tables* provided the base on which Isaac Newton and others would build. The star catalog, named for the Emperor and incorporating data of Tycho Brahe and others, contained 1,440 stars and was an invaluable aid to astronomers for nearly a century. Created out of a desire to demonstrate the perfect harmony and beauty of God's creation, Kepler's model of the solar system was not in the end particularly successful. But the theories regarding the movement of the planets which came out of it were, and his data proved essential to later scientists, including Newton when he developed his theories of specific gravity out of Kepler's work on the ecliptical paths of the planets. This curious diagram, more than the star charts he created or contributed to, exemplifies the process by which we ultimately came to understand the universal plan celestial cartography symbolically represents.

PLATE 37
Model of the universe
Johannes Kepler (Tübingen, 1596)
Engraving on paper, 8 x 8"
From *Mysterium Cosmographicum*

TABVLA III. ORBIVM PLANETARVM DIMENSIONES, ET DISTANTIAS PER QVINQVE REGVLARIA CORPORA GEOMETRICA EXHIBENS.

ILLVSTRISS⁰ PRINCIPI, AC DÑO. DÑO. FRIDERICO, DVCI WIRTENBERGICO, ET TECCIO, COMITI MONTIS BELGARVM, ETC. CONSECRATA.

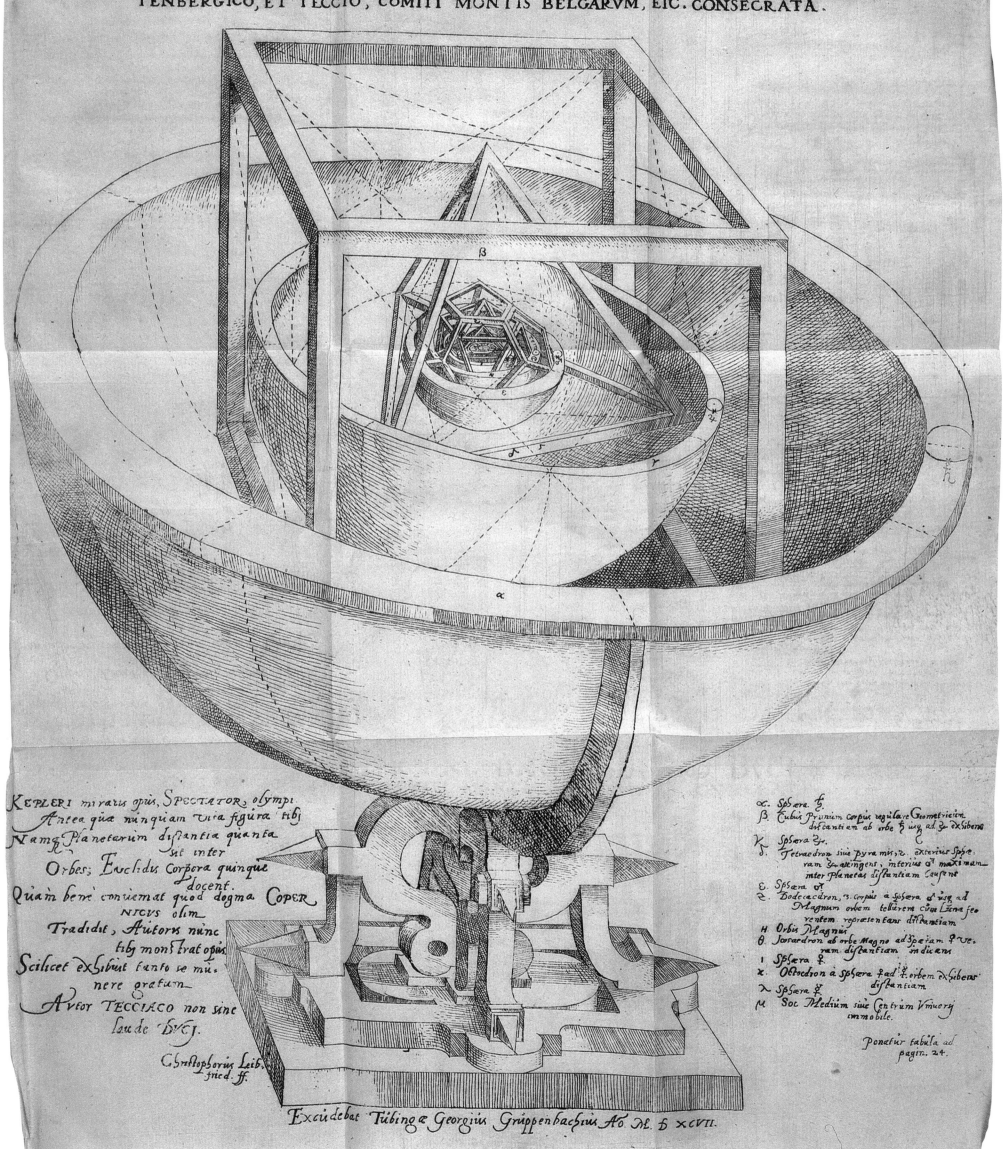

KEPLERI miraris opus, SPECTATOR, olympi.
Antea quàm nunquàm visa figura tibi.
Namq; Planetarum distantia quanta
  sit inter
Orbes, Euclidis Corpora quinque
  docent.
Quàm bene conuenat quod dogma COPER-
  NICVS olim
Tradidit, Autoris nunc
  tibi monstrat opus.
Scilicet exhibuit tanto se mu-
  nere gratum.
Autor TECCIACO non sine
  laude DVCI.

Christophorus Leib-
  fried. ff.

α. Sphæra ☿.
β. Cubus Primum corpus regulare Geometricum
    distantiam ab orbe ♃ usq ad ♄ exhibens
γ. Sphæra ♃.
δ. Tetraedron siue pyramis, ♃. exterius Sphæ-
    ram ♃ attingens, interius ♂ maximam
    inter Planetas distantiam causens
ε. Sphæra ♂.
ζ. Dodecaedron, ♂ corpus à sphæra ♂ usq ad
    Magnum orbem telluris cum Luna se-
    uentem repræsentans distantiam
η. Orbis Magnus
θ. Icosaedron ab orbe Magno ad Sphæram ♀ ue-
    ram distantiam indicans
ι. Sphæra ♀.
κ. Octaedron à Sphæra ♀ ad ☿ orbem exhibens
    distantiam
λ. Sphæra ☿.
μ. Sol Medium siue Centrum Vniuersi
    immobile.

Ponatur tabula ad
  pagin. 24.

Excudebat Tübingæ Georgius Gruppenbachius A° M. D XCVII.

Giovanni Paolo Gallucci (1538–1621?) is possibly best known for his works on astronomical and scientific instruments. He wrote one of the earliest texts on the magnetic needle as well as the first Venetian star atlas with coordinates, published in 1588. Interestingly, one of Gallucci's interests was astrology, and in addition to his more scientific writings he also wrote extensively on the subject of astrological predictions. In 1584, at the request of the bishop of Modena, Gallucci published a work on astrological medicine—using astrology to treat patients. The work was mostly borrowed from other authors, Gallucci acting as translator. However, in his *Theatrum Mundi* (*Theater of the Universe*), a work he produced in six volumes and dedicated to Pope Sixtus V, Gallucci pursued the subject of astrology in depth. In the introductory dedication the author

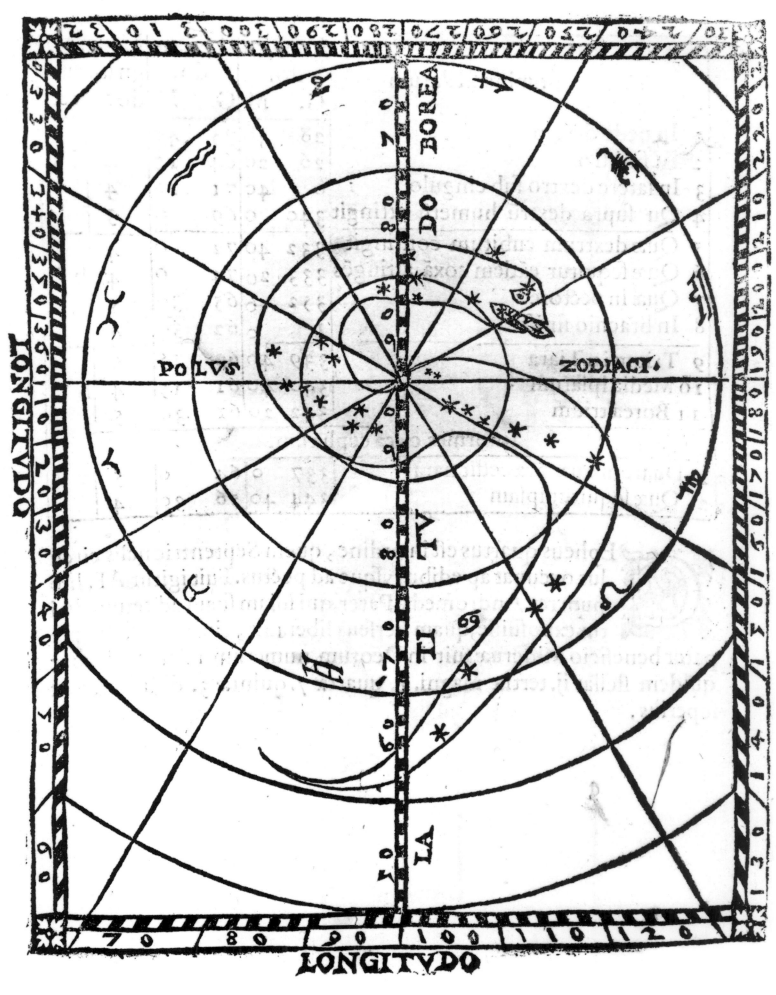

PLATE 38
Draco
Giovanni Paolo Gallucci (Venice, 1588)
Woodcut on paper, 6 x 5″
From *Theatrum Mundi*

encourages the pope to endow an astronomical observatory, and in the text of the work discusses the importance of astronomical observations for the advancement of astrological work. He discusses the natures of the individual planets, the various signs of the zodiac and their respective natures (i.e., masculine, feminine, commanding or obedient, etc.). Gallucci was convinced that astronomical observations of even the most cautious and conservative of scientists would reveal the effects the movement of the celestial bodies had on the nature of life on earth.

It is clear that Gallucci's maps included in the work were produced with scientific and astrological motivation; he perceived in his depiction of the constellations a meaning which went far beyond the images presented.

Gallucci's atlas, like Dürer's maps but unlike earlier

PLATE 39
Perseus
Giovanni Paolo Gallucci (Venice, 1588)
Woodcut on paper, 6 x 5″
From *Theatrum Mundi*

77

atlases, made it possible to locate constellations in the heavens. The coordinates used in the Gallucci maps are taken from Copernicus's *De Revolutionibus*. In Copernicus, the longitudes of the stars are measured from the first star of the constellation Aries rather than from the spring equinox. Interestingly, it was Copernicus who first accurately explained the phenomenon of the precession of the equinoxes as an alter-

ation of the plane of the earth's equator. Using the fixed stars, however, rather than the sun's apparent motion, as the basic reference point to determine star positions relegates the matter of precession to somewhere outside or beyond the celestial sphere, so to speak. The result is that the maps are "accurate" in relation to each other for all time, but do not reflect by their coordinates a relationship to the sun's movement.

PLATE 40
Andromeda
Giovanni Paolo Gallucci (Venice, 1588)
Woodcut on paper, 6 x 5″
From *Theatrum Mundi*

78

The forty-eight Ptolemaic constellations are represented in single maps, each accompanied by a catalog of the stars within the constellations indicating the magnitude of the various stars. Three of the twenty-one northern constellations are represented (plates 38, 39, 40), along with Gemini of the twelve zodiacal constellations (plate 41). Stylistically, the figures bear rather more resemblance to manuscript versions than the considerably more elaborate figures of Dürer or Honter. In these somewhat utilitarian and plain images, however, there is a certain charm. The little brothers Castor and Pollux forever together, Andromeda forever in her chains, the victorious Perseus, and the dragon coiled about the pole—each a bold yet naïve negative against the white sky.

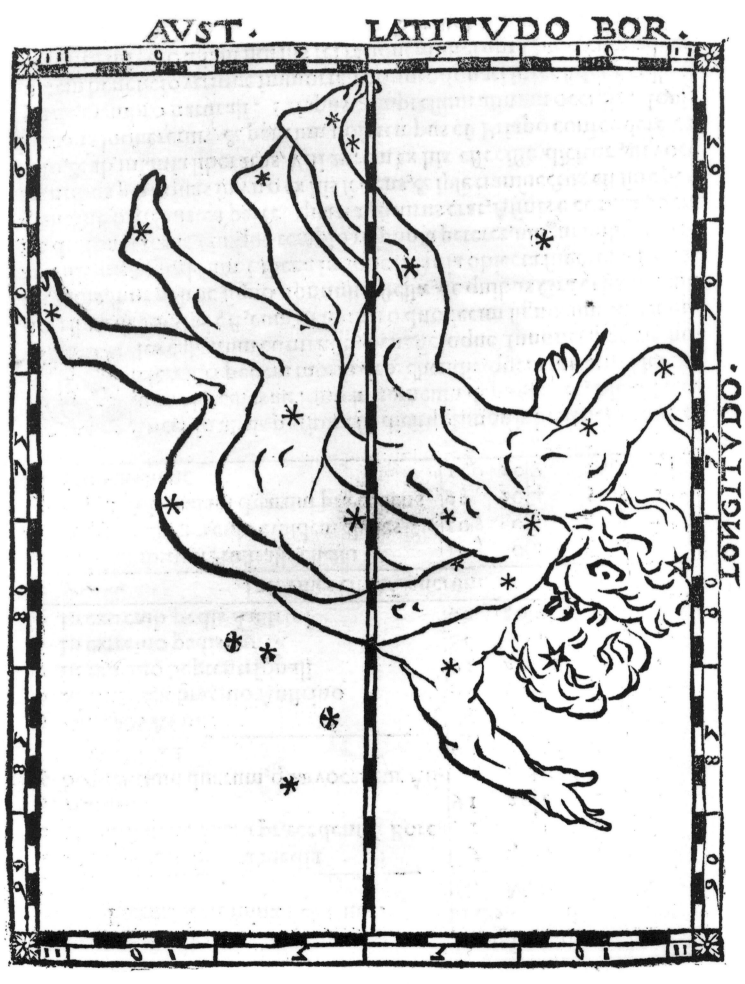

PLATE 41
Gemini
Giovanni Paolo Gallucci (Venice, 1588)
Woodcut on paper, 6 x 5″
From *Theatrum Mundi*

PLATE 42
Northern celestial hemisphere
Thomas Hood (London, 1590)
Engraving on paper, 12 x 12″

The success or value of a celestial chart depends upon the way in which the factual astronomical data is combined with the figural and visual representation of the constellations and stars. As with any kind of map, whether celestial or terrestrial, neither a purely visual depiction nor purely textual discussion is sufficient in itself; the two sources of information must be combined together to form a comprehensive and useful whole.

From the modern viewpoint, however, maps are sometimes judged from a purely aesthetic point of view, at times without regard to the data being presented. Since earlier examples of constellation depiction (see, for example, plate 5) were purely figural and did not supply actual coordinates for star positions, it is easy to lose sight of the celestial map as a depiction of astronomical data and tempting to judge it on purely artistic merits. For educational instruction or navigational purposes, though, the artistry of the celestial map is in many respects secondary to the way in which the creator of the work conveys information about the stars. Thomas Hood (fl. 1577–1596) was first and foremost a mathematician and lecturer on astronomy. Accompanying each constellation depicted (plates 42, 43) is an extensive text explaining the mythological background and citing the various names for each figure in different languages . Clearly, Lord Lumley, Hood's patron, whose coat of arms appears in the southern hemisphere, was interested in commissioning Hood to produce instructional tools which would be useful in astronomical

lectures. Artistically, however, the maps are somewhat disappointing. The tortured anatomical rendering and crude depictions of such figures as Canis major and Canis minor fall short in comparison to the fine artistic achievements of other celestial cartographers whose work is not only accurate but also beautiful.

Judging the merits of astronomical charts, however, unlike terrestrial mapping which demands accurate representation as an aid to navigation, an awkward depiction of the constellation of Capricorn can do no harm, provided the coordinates of the stars within the constellation are accurate.

The freedom enjoyed by the celestial cartographer in fact resulted in a variety of productions, exceeding in some ways the range of choices available to the terrestrial cartographer. That freedom, however, placed demands on the creator of maps of the heavens unlike those placed on the individuals who focused on the earth below. Great artists like Dürer or Cellarius could meet those demands, whereas others like Hood would struggle to convey the data of astronomers in an accurate as well as pleasing manner.

Hood's work is a reminder that there are many ways to present information concerning the heavens. What is perhaps lost in the shapes of the figures is compensated for by the explanatory text which accompanies them, filling a space with useful knowledge where in other cases fine line and aesthetic beauty would prevail.

PLATE 43
Southern celestial hemisphere
Thomas Hood (London, 1590)
Engraving on paper, 12 x 12″

PLATE 44
Celestial map
John Blagrave (London, 1596)
Engraving on paper, 10¼ x 10¼"
From *Astrolabium Uranicum Generale*

This complicated and busy looking creation (plate 44) says as much about John Blagrave's interest in scientific instruments and mathematical tables as it does about the heavens. Tables for calculating celestial movements and other phenomena fill its corners. The tracks of comets and even the elemental associations of the zodiacal signs—fire, air, earth, and water—crowd the map and seem at first sight almost overwhelming in their scope.

An inventor of scientific instruments, a mathematician and surveyor, Blagrave created this map to accompany his 1596 work on the Uranical astrolabe he invented. The map contains within its corners information and tables regarding the invention that are as informative as they are effective advertising for the instrument, which it seems, unfortunately, did not enjoy long-lived popularity. Blagrave's attempts with the astrolabe, however, were very much within the stream of effort exerted by contemporaries in the field: the desire to improve on existing instruments used to measure distance by citing celestial reference points motivated astronomers and scientists like Blagrave to develop variations on the astrolabe and other instruments. Literally "star-taking," the astrolabe was a tool to measure distances taken from the stars; its value to astronomy is revealed in its connection with Blagrave's

map, for although the Uranical astrolabe did not survive, Blagrave's map did.

In addition to the information on the zodiacal signs and celestial motion tables, Blagrave includes the "Cometa of 1596," shown at the hind feet of Ursa major, and a note regarding the "Stella Nova" in Serpentarius. Along with the royal coat of arms and motto in the upper right-hand corner, Blagrave depicts the "Corona australis" (southern crown) as the "Corona anglie"—a nice touch of patriotism on his part. Nationalism and the dedication of constellations to royal and noble individuals were not unusual features of star charts, nor was the appearance (and disappearance) of constellations serving special interests. The astronomer J. J. L. de Lande would later on create a short-lived constellation in honor of his cat, Felis.

Represented here are two star groups first raised to the status of constellations by Tycho Brahe: Coma Berenices and Antinous. Ptolemy had designated these groups as clusters of unformed stars, but it was Tycho Brahe who first represented them as, respectively, the crowned head of hair (as the image is usually shown) of the third-century Berenice, queen of Ptolemy Eurgetes (a recompense for a lock of her hair that was stolen from a temple dedicated to Venus), and the young boy Antinous, boy-lover of the Roman emperor Hadrian.

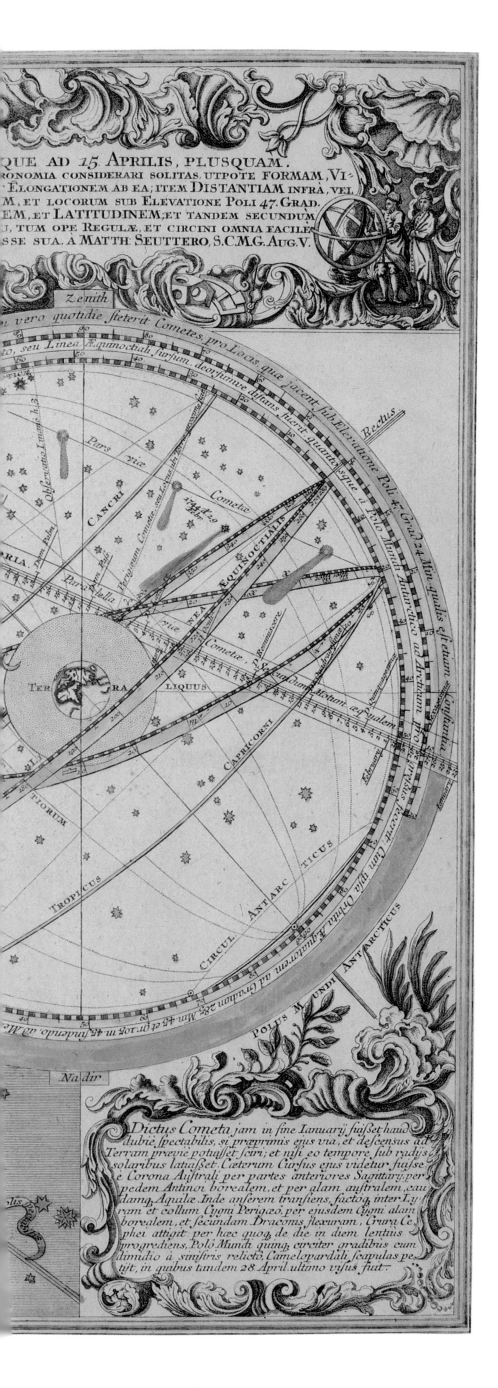

Comets—from the Greek for long-haired, suggested by the tail of light—belong to a class of nebulous bodies within the celestial system that behave quite unlike other heavenly bodies. Varying in brightness, orbit eccentricity, and shape of tail, comets seemed to early observers some sort of alien and perhaps miraculous intruder into the orderly system of the heavenly firmament. It was noted that the tails of comets appeared to point away from the sun, and that in some cases the proximity of planets such as Jupiter appeared to affect a comet's path through the sky. Moreover, certain comets, the most famous being Halley's comet, seemed to appear regularly and periodically.

The association of comets with human events has its own fascinating history. The comet of 1456, later named for English astronomer Edmund Halley, was believed to announce the threat of the Turkish Empire to Europe. Two centuries later, a Spanish writer would claim its appearance in 1682 as heralding the triumph of the Catholic faith. The comet of 1695 was interpreted by another writer as a sign to Christian princes to join ranks and regain the Holy Sepulchre, while by the time of the comet of 1727, Miguel Martinez y Salafranca would write a witty argument against the theological interpretation of the appearance of comets—where, he asks, were the comets when the Great Flood was imminent? Well into the nineteenth century, however, comets were still seen and interpreted as portents of impending disaster, and less than 100 years ago, the coming of Halley's comet was believed by some to announce the end of the world.

With the invention of the telescope, many more comets could be observed in the sky than were visible to the naked eye. On page 88 the first published telescopic observation of a comet can be seen. Johann Baptist Cysat (ff. 1600), a Jesuit astronomer, shows that the orbit of the comet is parabolic rather than circular. Here it is seen passing through the constellations of Boötes and Ursa major from December 8 to 9. The work also includes Cysat's discovery of the Orion nebula and two of Saturn's satellites (not shown).

In plate 45, Matthew Seutter (1678–1756), a German cartographer and instrument maker, depicts the path

PLATE 45
The comet of 1742
Matthew Seutter (Augsburg, c. 1745)
Copperplate engraving, hand colored, 18½ x 24″
From *Atlas Novus sive Tabulae Geographicae*

of the comet of 1742 from March 13 to April 28, showing the length and direction of the tail. In addition he includes the paths of the comets of 1618–19 (as recorded by Kepler), 1642, and 1683.

By Seutter's time, significant advances had been made in theories regarding comets. Halley had, in 1682, observed that the comet of that year bore a striking resemblance to those which had appeared in 1531 and 1607, and he concluded that all were the same—the comet seemed to reappear approximately every seventy-six years. He predicted its return in 1757, and though it did not appear until 1759, Halley's name was given to the comet, and it was recorded again in 1835 and 1910. Well over a dozen periodic comets were recorded by the dawn of the twentieth century. Sophisticated instruments would explore the farthest reaches of space, and photographic studies of the comets' paths would replace the simple depictions of earlier centuries.

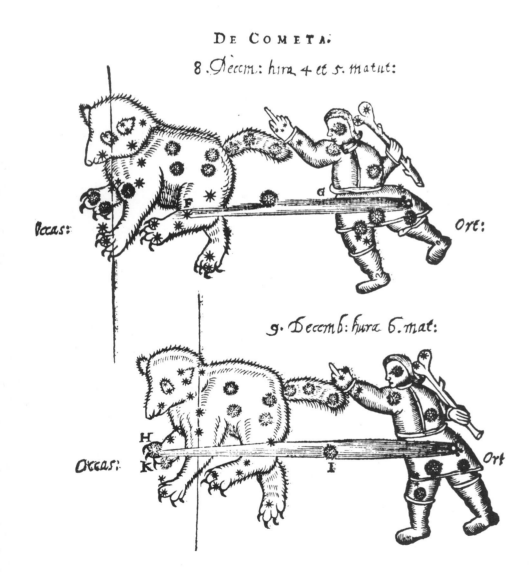

Among the many contributions of Christian Huygens (1629–1695), Dutch natural philosopher and scientist, was the construction of a long-focus refracting telescope with which he made a number of discoveries. Plate 47 presents Huygens's discovery of the Orion nebula, which had also been sighted by Cysat in 1618, unknown to Huygens. In Huygens's book, *Systema Saturnium*, his discoveries of the satellite Titan and the ring around Saturn are also presented.

A nebula is an indistinct cloud of distant stars, in this case surrounding the constellation of Orion. Huygens described the phenomenon as a view through an opening in the dark heavens which revealed a brighter region farther away. The Orion nebula can, in fact, be faintly seen with the naked eye. With sufficient telescopic power, the nebula appears to resolve into a cluster of tiny stars, although many nebulae do not resolve and would appear even through a powerful telescope as clouds rather than as star clusters.

In 1771, the French astronomer Charles Messier recorded 103 nebulae. That number was considerably increased in the nineteenth century by other astronomers with more modern instruments. They also instituted the classification of nebulae into structural types, and posited a more refined definition of their make-up. It was clear in Huygens's time, however, that the telescope would reveal a more complex and intricate system in the celestial sphere than could have been imagined previously.

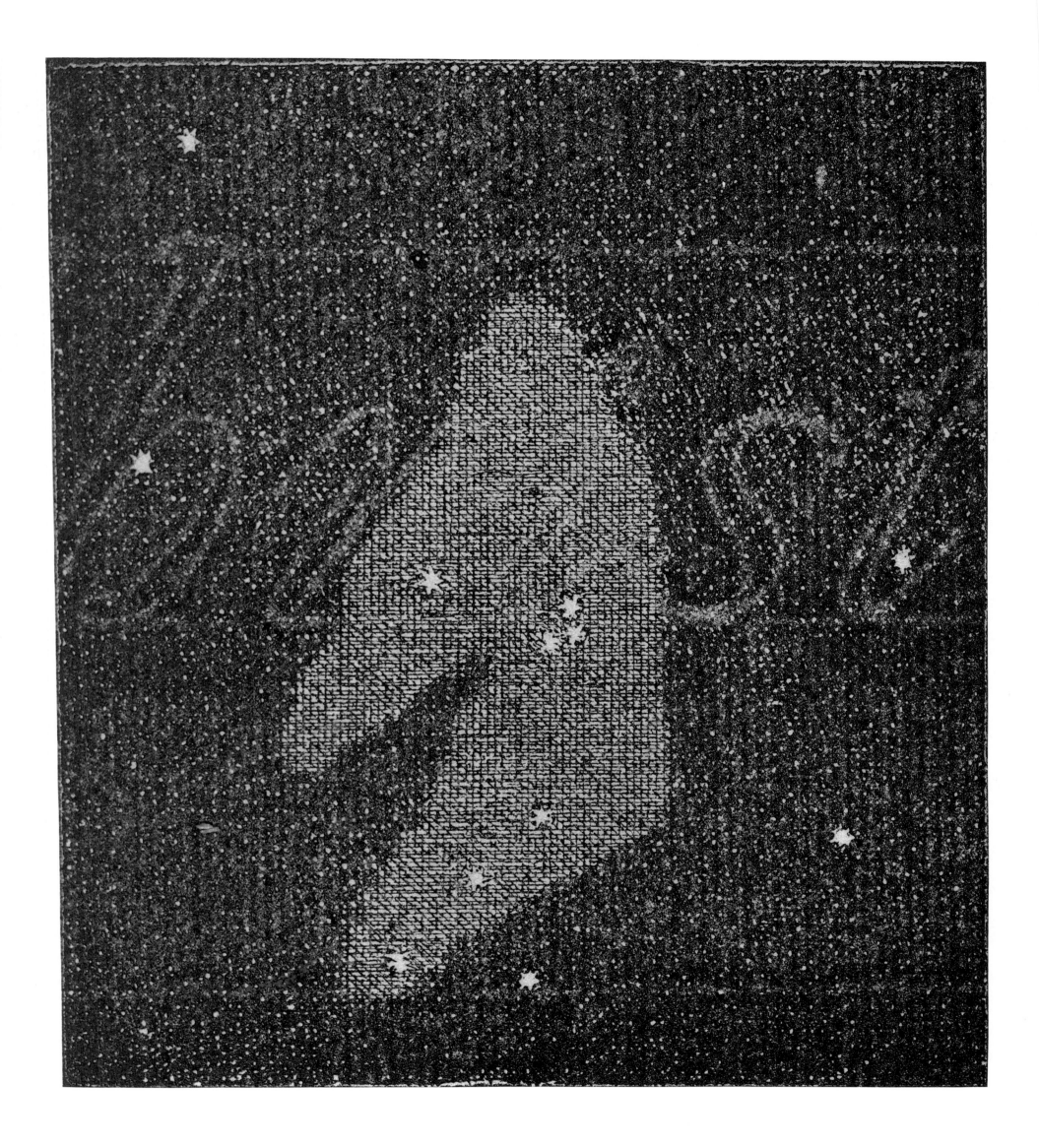

PLATE 47
Chart of the Orion nebula
Christian Huygens (The Hague, 1659)
Engraving on paper, 6 x 5″
From *Systema Saturnium*

DESIGNING THE HEAVENS

PLATE 48
Cassiopeia
Johann Bayer (Augsburg, 1603)
Engraving on paper, 14 x 16″
From *Uranometria*

Johann Bayer's (1572–1625) celestial atlas, *Urano-metria* (named after the kingdom of the Muse of Astronomy, Urania), first published in Augsburg in 1603, introduced a feature that would not come into widespread use in charts of the heavens until the middle of the eighteenth century: the identification of astral magnitude (brightness) with a lettering system. Greek letters are used for the brighter stars and Roman letters for the fainter, the order of the letters corresponding, generally speaking, to the decreasing brightness, as can be seen in this chart of Cassiopeia (plate 48). At first, it seems, astronomers did not realize the value of Bayer's system. Although the Italian cartographer Alessandro Piccolomini had earlier used a somewhat similar system, it was not until Augustin Royer used the Bayer letters in 1679, followed shortly by John Flamsteed, that the system gained currency among celestial chartmakers.

Bayer's atlas also adds twelve new constellations, in the southern sky, to the forty-eight of Ptolemy. These had been charted not long before by the Dutch navigator Pieter Dirckszoon Keyzer (known as Petrus Theodori) from Emden, who had corrected the sightings of Amerigo Vespucci and Andrea Corsali as well as information supplied by Pedro de Medina. Keyzer's constellations, including the bird of paradise, the peacock, the phoenix, the flying fish, and the toucan, along with other exotic creations, appear in the forty-ninth plate of Bayer's atlas (not shown). The system of lettering is absent, however, and was not applied until the eighteenth century by the French astronomer Abbé Nicolas Louis de Lacaille.

Bayer's stellar lettering system, which we still use for stars visible to the naked eye, and his presentation of Keyzer's constellations were significant contributions to celestial cartography. Ironically, it may be that his work on the atlas had an ulterior motive. Bayer, by profession a lawyer, was really an amateur astronomer. He dedicated his atlas to the city council and to two leading citizens of Augsburg, who rewarded him with an honorarium and, later, a seat on the council as legal adviser.

The atlas was, nonetheless, a popular success. But its innovative system also presented difficulties, for unlike his predecessors Bayer chose to label his figures in reverse. What other astronomers had called the right side of the constellation, Bayer labeled the left side. The reason for this reversal is unclear. Toward the end of his life, Bayer assisted Julius Schiller in the preparation of that astronomer's star atlas [see pages 96–99].

Representing a three-dimensional schema on a flat surface was a problem for cartographers of heaven and earth that gave rise to a tremendous variety of solutions. Wilhelm Schickard (1592–1635) proposed the system represented here (plates 49, 50) to depict the constellations. It was later copied but never universally accepted.

The northern and southern hemispheres appear as cones that have been split at the summer solstice. Joined together they would form a double-conical version of the globe. Because the images of the sky are projected as seen from the earth, the constellations would appear *inside* the cones.

Schickard's charts also include a popular feature of the time that would be further developed by Julius Schiller, with whom Schickard collaborated. Schickard's constellations are given, in addition to their traditional nomenclature, biblical references, reflecting the practice of "Christianizing" the heavens in the late sixteenth and early seventeenth centuries. The connections made between classical and Christian allusions demonstrate a considerable amount of dexterity. For example, the reference given for Cassiopeia is 1 Kings 2:19. In Greek mythology Cassiopeia is the wife of Cepheus, mother of Andromeda; the biblical allusion is to Bathsheba, wife of King David and mother of

PLATE 49
Northern hemisphere
Wilhelm Schickard (Nördlingen, 1655)
Engraving on paper, 5½ x 5½"

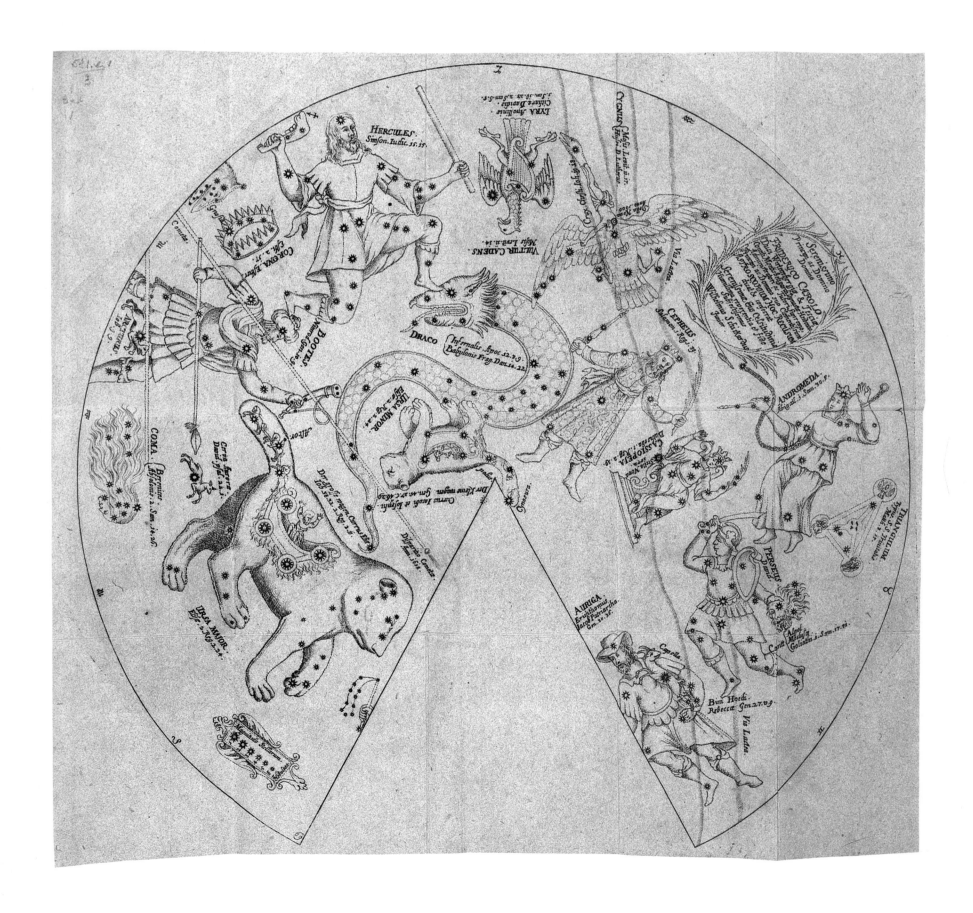

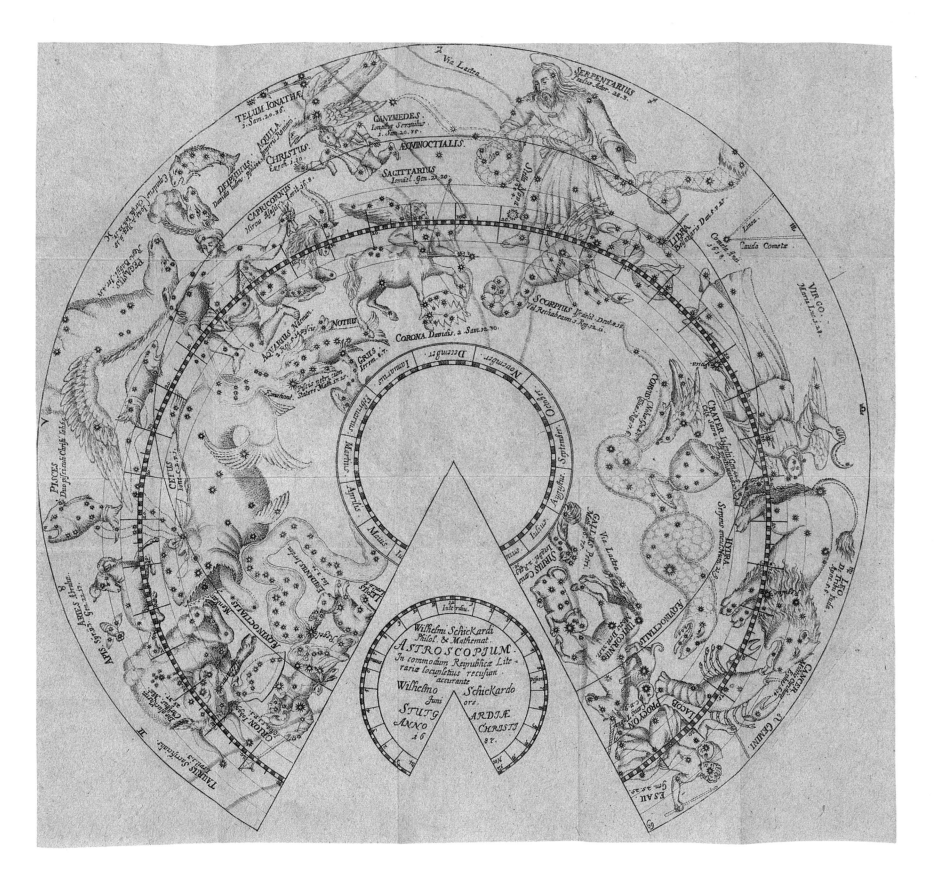

PLATE 50
Southern hemisphere
Wilhelm Schickard (Nördlingen, 1655)
Engraving on paper, 5½ x 5½"

Solomon: "So Bathsheba went in to King Solomon to speak for Adonijah. The king rose to meet her and kissed her, and seated himself on his throne. A throne was set for the king's mother and she sat at his right hand, . . ." and with a sleight of hand Schickard converts the seated Cassiopeia into the enthroned Bathsheba. Likewise, Perseus is compared to David, and therefore Medusa to Goliath. Virgo's obvious counterpart is the Virgin Mary.

Schickard's citations of novas (new stars) and comets on his charts include the "Stella Nova 1572" in Cassiopeia and the comet of 1618, shown with its trail through the constellation Boötes and the tail of Draco. Peter Apian had been the first, in his 1540 *Astronomicon Caesareum*, to resurrect the Arabic reference (mentioned by al-Sufi) to the bright stars within Ursa major being depicted as a four-wheeled wagon pulled by horses. Here, Schickard incorporates Apian's image in the belly of his Big Bear and echoes it in a smaller version in the Little Bear. In addition, he provides biblical references to chariots and wagons, thus appropriating a celestial image born in Islam and used in Europe.

The work of Julius Schiller (?–1627), a devout German Catholic, reflects an extraordinary interaction of faith and imagination in the interpretation of the heavens. His atlas, the *Coelum Stellatum Christianum*, (*Christian Celestial Heaven*), published in 1627, was the result of his efforts to completely rename and redraw the celestial constellations as they were known in his time, to convert the sky into a system of Christian symbolism in which every constellation is given a place within biblical iconography. At the same time, Schiller's book presented the most up-to-date astronomical information.

More than the passing fancy of a religious fanatic, Schiller's atlas was the result of arduous and careful preparation. Many of the noted contemporary astronomers of the time—Tycho Brahe, Johannes Bayer, and Johannes Kepler among them—contributed to Schiller's work and advised him on observational data. The result is an accurate depiction of the star system in a radical new form (see plates 54, 55). The zodiac signs are transformed into the twelve apostles, while the constellations of the northern and southern hemispheres become images and personages from the New and Old Testaments, respectively. Latin letters are used to identify the coordinates of the major stars.

One can imagine Schiller looking at the celestial maps of the day and thinking to himself how cluttered the heavens were getting, how inconsistent the iconography, how confused and archaic the mixed symbolic metaphors looked, pagan image next to modern discovery—a sky full of strange, half-forgotten figures from ancient myths. How much better, and instructive, a universal heavenly system would be.

The constellation Draco here (plate 51) has been reinterpreted as the Massacre of the Innocents,

PLATE 51
Massacre of the Innocents (alias Draco)
Julius Schiller (Augsburg, 1627)
Engraving on paper, 10½ x 13″
From *Coelum Stellatum Christianum*

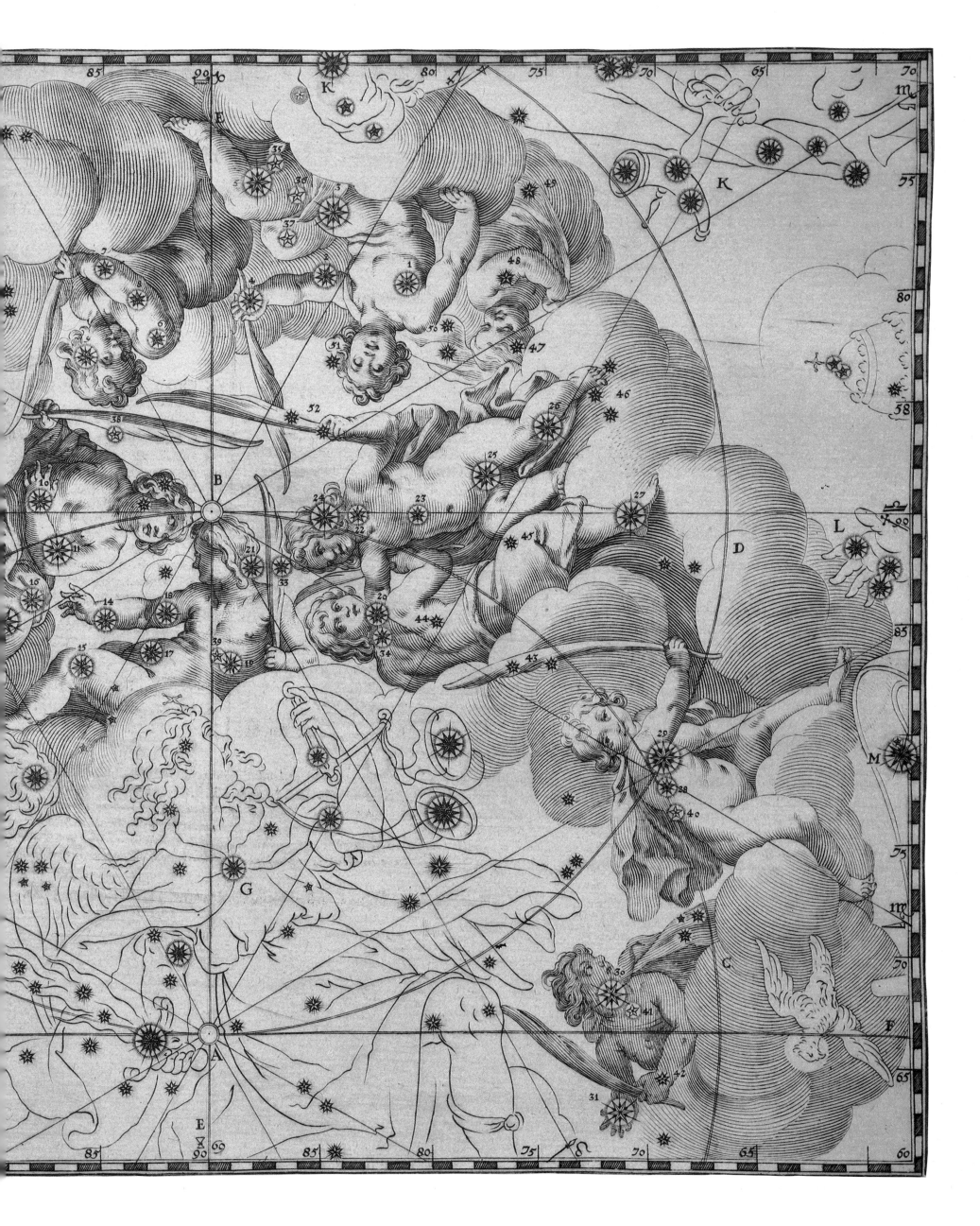

PLATE 52
The Three Magi (alias Hercules)
Julius Schiller (Augsburg, 1627)
Engraving on paper, 10½ x 13″
From *Coelum Stellatum Christianum*

Hercules (plate 52) as the Three Magi, and Pegasus (plate 53) as the Archangel Gabriel. Cellarius's *Atlas Coelestis*, published later in the century, expanded on the system as a whole to depict the heavens as no one before had presented them. These maps, in fact, gained Schiller greater popularity than his own work.

For obvious reasons, Schiller's atlas never received universal approval or acceptance, although it did enjoy a certain amount of popularity. It is more than a curiosity, however. Schiller's charts of the heavens reveal that even in the early seventeenth century the seeds had been planted for a reassessment of celestial cartography.

His example was echoed later, in the modern period, when nonrepresentational charts came to convey data without images. The confusion of a mixed system would give way to our clean and "scientific" charts of star locations and, in the present day, to satellite photographs. And yet, just as Schiller was unable to erase the images that had come into existence over a thousand years before him, the ancient constellations still have not become extinct, even after they have vanished from celestial charting. Several biblical images also remain in common use, including the cross, Noah's dove, and the unicorn (Monoceros).

PLATE 53
The Archangel Gabriel (alias Pegasus)
Julius Schiller (Augsburg, 1627)
Engraving on paper, 10½ x 13″
From *Coelum Stellatum Christianum*

COELI
CHRISTI
SPHÆRIUM

STELLATI
ANI HAEMI
POSTERIUS.

PLATE 54
Schiller's Christianized Heaven,
southern hemisphere
Andreas Cellarius (Amsterdam, 1660)
Copperplate engraving, hand colored,
18½ x 22"
From *Atlas Coelestis seu Harmonia
Macrocosmica*

STELLAT[A]
[F]ANI HÆ[MI]
RIUM PRIUS.

PLATE 55
Schiller's Christianized Heaven,
northern hemisphere
Andreas Cellarius (Amsterdam, 1660)
Copperplate engraving, hand colored,
18½ x 22″
From *Atlas Coelestis seu Harmonia
Macrocosmica*

PLATE 56
Thai manuscript of the universe, detail

Even before astronomers like Schiller sought to reshape and redraw the constellations, it was clear that imagination, guided by cultural and religious background, shaped what one found in the sky. The Babylonians, the Egyptians, the Greeks, and Europeans each presented their gods and other images in fanciful and meaningful arrangements of stars.

Oriental images of the universe, however, are virtually unintelligible to most Westerners, bearing little resemblance to the celestial cartography that had developed predominately out of the Judeo-Christian tradition. The Thai conception of the universe developed out of a combination of Asian and Oriental influences—religious and philosophical—that resulted in a cosmography far removed from anything developed in the West. Though some scholars have attempted to see a common goal toward which both the West and East strived in astronomical study, the divergent motives underlying the two worlds' approach to that scientific inquiry make the similarities, viewed in retrospect, somewhat less significant than is sometimes suggested.

Unlike the Western approach to science, with its abiding faith in logic and the value of empirical analysis, the Eastern approach is distinguished by a curiosity in the illogical, irregular, and the fantastic. The Platonic belief in eternal patterns combined in the West with faith in a perfect creation, forming the base of scientific study. In the East, however, other forces led to a faith in the dynamic vitality of a universe in which constancy and logic could not be expected, and the complexity of nature was such that no simple system could be deduced to explain it. To the Eastern mind, the world was far too evanescent and subtle to be categorized. Confucian teaching, with its emphasis on moral and social behavior, led to a historical approach to the phenomena of earth and sky, and the records of social and celestial activity which were compiled proclaimed the quixotic and irregular vitality of man and universe.

With the influence of Buddhism, the Thai conception of the universe took on yet another feature distinct from the West. More than a depiction of the cosmos, the Thai version was also an allegorical representation of the life of Buddha. The images were intended as "reminders by association" (*paribhogacetiya*) of the major events in Buddha's life and were meant to recall the elements of one's religion.

Just as the actual *stupa*, or temple, is a reminder to

Buddhists and Hindus alike of *Meru*—the central mountain of the Indian world—so the image of the *stupa* here (plate 56) is intended to recall the pavilion of Indra at the summit of *Meru*, the spire pointing to the heaven of Brahmā and the more abstract heavens of the blessed beyond. The spire is also symbolic of the axis of the universe, reaching down through the base of the *stupa* into the earth and the waters, guarded all around by various creatures, monsters, and spirits, and through the world, represented by the intertwined elephants.

The tree form of the ribbonlike ties (plate 57) between celestial and terrestrial worlds can be seen as an image of the Bo tree which sheltered Buddha at the moment of his enlightenment, and the whole of the depiction can be seen as a tool for seeking to visualize Buddha.

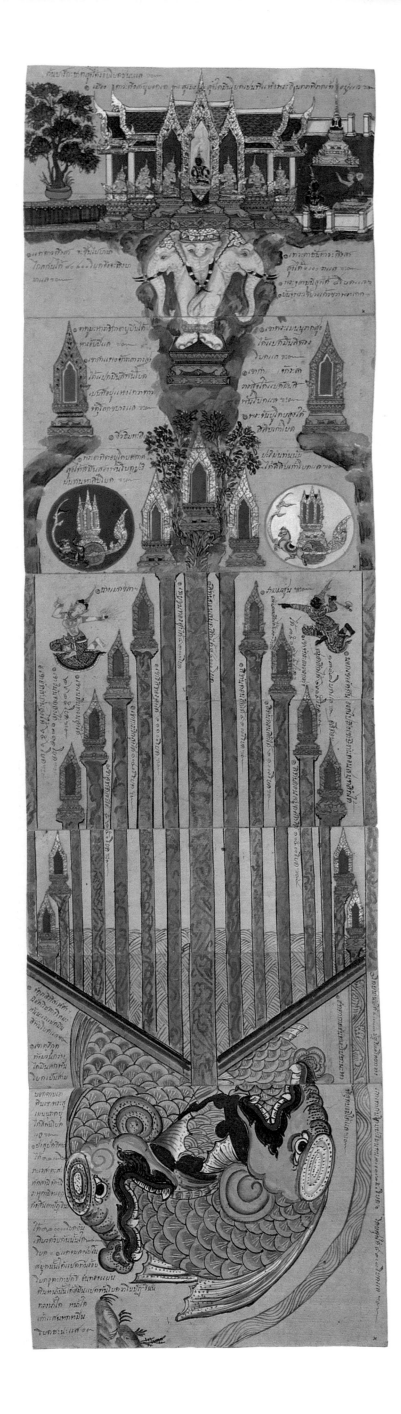

But to Buddhists and non-Buddhists alike, the images are open to various interpretations. The complexity and variety of the images delighting the viewer—the interconnected spiritual and physical worlds blurring distinction between the tangible and the invisible—obscure any simple or logical explanation.

This interconnectedness between celestial and terrestrial spheres, so unlike the structures of Western cosmography, underscores the importance of religious belief as a factor in cartography. In Thai culture, as indeed in many Eastern religious traditions, the spirit world and the physical world operate together in a far more tangible fashion than is generally acceptable in the West. No Thai home is complete without a "spirit house" located on the ground, or nearby, for occupation by household spirits.

Prepared on folding panels of paper made from the bark of the Koi tree, this Thai schema presents the world of man and spirit, in a series of powerful images—the steep-pitched roofs of the spirit dwellings and the elephants (symbols of royalty and power) (plate 57). No star coordinates are provided, for within the framework established none are necessary: this is a celestial chart which seeks to find relationships and meanings not found in Ptolemy's *Almagest* or in Kepler's system of regular polyhedrons.

The Thai chart, however, possesses a unity of design

PLATE 57
Thai manuscript of the universe
(Thailand, 19th century)
pen and ink, watercolor on Koi paper;
accordion-folded sheets, 14½ x 5½″

106

PLATE 58
Thai manuscript of the universe, detail

and cohesiveness of symbolism which is often lacking in the Western system, the orderly, colorful system of temples, streamers, and spirits as compared to the motley arrangement of Greek mythological figures with odd and exotic animals and configurations. How stately and dignified the progression of the Thai work next to a chart by Hood or Blagrave. Perhaps more important than any comparison, though, the Thai chart of the universe reminds us that ultimately the sky above is, unlike the earthly domain, the property of all and thus open to the interpretation of any who gaze heavenward. Whether peopled by ancient gods or the discoveries of the modern age, the heavens offered throughout the age of celestial cartography an enormous canvas on which to paint—a kind of mirror image of life as those who lived below came to know it.

PLATE 59
Northern hemisphere, detail
Joan Blaeu (Amsterdam, 1648)
Engraving on paper, 14 x 14″

These details from a large-scale map (plates 59, 60) of the world by the famous Blaeu family firm in Amsterdam indicate the qualities that made the Dutch the masters of cartography in the seventeenth and eighteenth centuries.

The two celestial hemispheres occupy the upper corners of the map. Although relatively small (thirty-five centimeters in diameter), the insets present the constellations clearly and accurately, including the Ptolemaic constellations as well as those discovered by Keyzer and first recorded by Johann Bayer. It is not surprising, however, that these maps—by Joan Blaeu, son of Willem Janszoon Blaeu (1571–1638) the firm's

founder—should be so fine. Willem had worked with Tycho Brahe at the Hven observatory. As a result of his interest in astronomy and navigation, he had left his work as a clerk in his cousin's mercantile office to open a business in globes and maps that would grow to be one of the most important in Europe. In 1602 he produced his first celestial globe, on which this work is certainly based.

The inclusion of celestial charts on terrestrial maps was to become a standard practice of the period. As navigation outgrew the confines of the Mediterranean and nations competed for new ports and trade routes in the Atlantic, the need for an accurate and reasonable

PLATE 60
Southern hemisphere, detail
Joan Blaeu (Amsterdam, 1648)
Engraving on paper, 14 x 14″

means of determining geographic longitude became more pressing. The hope that celestial charts would provide those means strengthened the natural association between the two realms of mapping. As astronomers had been the ones to define the parallels of the earth by studying the apparent movements of the sun and stars, so too might they provide a system to determine a ship's position at sea by the position of the stars. How appropriate, then, the inclusion of the constellation Colomba Noe (Columba Noachi) near the prow of Argo-Navis in the southern hemisphere. Also present is the dove Noah sent out to find land after the Flood.

109

Part of a larger celestial work, this set of gores (plate 61)—the triangularlike sections into which a globe can be separated on a flat page—reveals the quality and beauty of the work of Willem Blaeu, the inspiration for the preceding work, and the culmination of not only the most recent astronomical observations but the finest stylistic presentation of the time. The gore was one of several solutions developed to depict the celestial sphere just as the system had been used to depict a terrestrial globe. The difficulty of presenting on a flat surface the curving space of the sky, or the curving space of land, was conveniently solved by this globe form and served as an inspiration for future work by other astronomers and artists throughout the century. As late as the eighteenth century and even the nineteenth century, pairs of globes—one terrestrial and one celestial— would be sold in the shops of mapmakers in Europe, England, and America.

As a result of an influence by earlier and antique work, and part original genius, Blaeu's work set new standards for artistic efforts, and was widely imitated in the seventeenth century. Not content with only scientific accuracy (Blaeu revised his work to incorporate recent discoveries in later editions), the author is credited with developing the depictions of the constellations represented here, creating a style for celestial charts that—though in some ways reminiscent of earlier work—is in the final product a new and recognizably original effort.

The work of Willem Blaeu enjoyed considerable popularity at a time when the Dutch were expanding terrestrial exploration, sending its fleet into Asian, African, American, and Arctic waters, and widening the realm of the known world. Toward the end of his life Blaeu became the official cartographer to the Dutch East India Company, thereafter succeeded by his sons who would claim superiority in virtually all realms of cartographic production. Truly in the grandest of traditions, a combination of precision and aesthetic accomplishment resulted in work that both satisfied the desire for beauty and the needs of those engaged in the activities of commercial enterprise.

Joseph Moxon's map for his book *A Tutor to Astronomy and Geography* (London, 1674), goes further than Blaeu's in actually placing terrestrial and celestial hemispheres side by side, indicating the more "universal" approach man was taking in the late seventeenth century.

The two spheres (plate 62), both revolving on the north equatorial pole, meet at a single point at the centerfold of the map where it is mounted in Moxon's book. As the book opens, the two spheres are "separated," and as it closes the celestial dome merges with the earthly world below. The viewer can, vicariously, imagine the stars from anywhere in the world, although, in fact, the geography of the world was still imperfectly understood, as is indicated by the depiction of California as an island.

Perhaps purposefully, Moxon engraved his celestial chart in the stippling technique, the process of engraving with small dots instead of lines, giving the constellations a shadowy, ghostly glow. On the other hand, he surrounds his spheres with "timely" vignettes of universal importance: scenes from the Old and New Testaments, the ever-present sun shining on an angelic earth reclining on the soft folds of billowing cloud formations. Repeated throughout is Hebrew lettering, which perhaps to Moxon suggested things ancient and a complete tying together of the universe. Unfortunately, Moxon's Hebrew is as inaccurate as his depiction of California, reflecting the constant temptation in mapmaking, and other pursuits, to exceed the limits of one's own knowledge in order to realize a vision of truth.

PLATE 62
Celestial and terrestrial spheres
Joseph Moxon (London, 1674)
Engraving on paper, 18½ x 22"

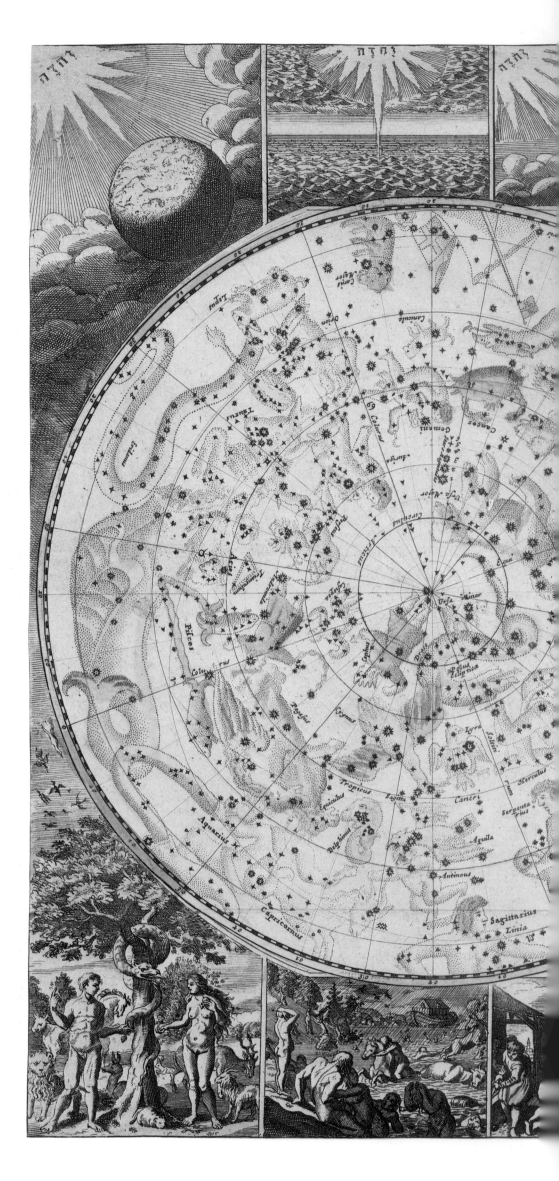

LONDON.
Printed by J. Moxon. And are to be sold at
The signe of Atlas. In Warwick lane

To the most Reverend Father in God
JOHN, Lord Arch Bishop of Canterbury, His Grace Primate of all England and Metropolitan &c
This Mapp is humbly D.D. by J. Moxon.

UM STELLAT CVM SUBIECTO TERRESTRI.

PLATE 63
Northern hemisphere
Andreas Cellarius (Amsterdam, 1660)
Engraving on paper, hand colored, 18½ x 22"
From *Atlas Coelestis seu Harmonia Macrocosmica*

Very little is actually known about the creator of the most beautiful celestial atlas ever made. All that we know for certain is that Andreas Cellarius was rector of the Latin School at Hoorn in Holland when his book of twenty-nine maps, the *Atlas Coelestis seu Harmonia Macrocosmica* (*Celestial Atlas of Universal Harmony*), first appeared in 1660. His life is otherwise a mystery, although there is some evidence that he may have previously produced an atlas of Poland.

The *Atlas Coelestis* contains maps of the Ptolemaic universe, the Copernican system, and the Tycho Brahe compromise between the two systems. The remaining maps are variations on these possibilities, with different kinds of constellation representation (notably Schiller's). The title page, of the same quality and style as the maps, depicts a group of astronomers in various styles of dress grouped about Urania, Muse of Astronomy.

The two plates reproduced here (plates 63, 64) show the northern and southern celestial hemispheres from the equatorial poles at angles to both the ecliptic and the equator. The earth is inscribed within the overall depiction, giving the viewer the illusion of looking through the celestial realm at the earth delineated below. Skillful handling of the engraving techniques furthers the spatial illusion, and the masterfully shaded and colored figures appear to float in the heavens above the earth.

In the border, the sky is shaded in almost surrealistic sunset colors. Putti carouse in the clouds, some supporting banners carrying the titles of the maps, others seeming to peer over the edge into the vast spaces of the hemispheres. Below, astronomers with their pupils study, engage in discourse, or manipulate various astronomical instruments.

The depictions of Ptolemaic constellations combine various traditions and yet maintain a consistent overall unity. Serpentarius in classical undress seems to blend with the figure of Boötes, dressed for a northern European winter's day in fur hat, heavy boots, and cloak. Likewise, Argo, as a seventeenth-century ship in full rigging, sails calmly by Orion, dressed in the gear of a Roman centurion. The stars, heightened with gold, create an appropriately dazzling effect.

The non-Ptolemaic constellations depicted are those of Tycho Brahe (Coma Berenices and Antinous) and Petrus Plancius (Columbia Noachi, Camelopardalis, Monoceros, the rivers Jordanus and Tigris, among others), as well as Keyzer (Indus, Pavo, and Piscis Volans).

116

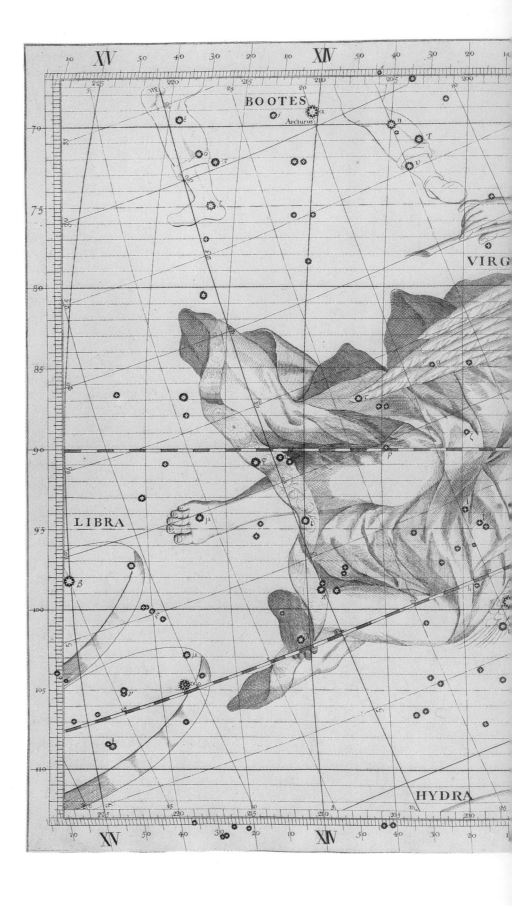

The first Astronomer Royal of the Greenwich Observatory, John Flamsteed (1646–1719), was so dedicated to his work and its perfection and accuracy that he not only died before it could be published, but was also accused in his lifetime of hoarding his data for himself. When, apparently out of frustration or impatience, his contemporaries Isaac Newton and Edmund Halley attempted to publish a portion of his first findings, Flamsteed had the work suppressed, insisting that his observations were not sufficiently accurate for publication. Understandably, his relationship with his fellow scientists became strained.

The founding of the Greenwich Observatory came out of a need, in the mid-seventeenth century, for accurate telescopic observations of the heavens. The urgency felt by Newton, Halley, and others came from a belief that such observations could be used to solve the most serious and pressing problem facing navigators—the determination of longitude. Part of the hope that England would be first to realize a solution was Flamsteed's appointment in 1675, thus the disappointment of his colleagues when he did not produce.

Published posthumously as the third part of the *Historia Coelestis Britannicae* (*British Celestial Charts*) in 1725, Flamsteed's *Stellarum Inerrantium Catalogus Britannicus* (*The British Star Catalogue*) describes 2,935 stars observed from Greenwich, exceeding all former catalogs of stars in number and accuracy. It

PLATE 65
Virgo
John Flamsteed (London, 1729)
Engraving on paper, 24 x 20″
From *Atlas Coelestis*

was also the first telescopic catalog of the northern stars. In addition, Flamsteed prepared a set of celestial maps of the twenty-five constellations visible from the Greenwich Observatory to accompany his catalog.

These were published in 1729, ten years after his death, with the title *Atlas Coelestis* (not to be confused with Cellarius's atlas of the same name). Plate 65 shows the constellation Virgo, with coordinates for 1690, and Bayer's Greek letters identifying the brighter stars.

The figures in the plates are from drawings by Sir James Thornhill and typify the classical tastes of the period. Sir James, an artist of fame and in much favor at the time, had been commissioned to decorate the dome of Saint Paul's Cathedral with painting (much

against the wishes of the architect Sir Christopher Wren), and one cannot help wondering whether working within the celestial heights of that dome had any bearing on his designs for Flamsteed's maps. His sketchbooks, however, one of which is in the British Museum, indicate that Thornhill kept handy a ready portfolio of allegorical and mythological subjects should he be commissioned to decorate a wall or ceiling.

In retrospect it might be said that Flamsteed's insistence on perfection paid off in the end; his *Atlas Coelestis* became the most important star atlas of the eighteenth century.

If lawyers and poets and teachers could produce charts of the heavens, so too could an architect. The map shown here is the work of one of Louis XIV's royal architects, Augustin Royer (fl. 1680?–1700?), published in Paris in 1679. That Royer was not the principle architect responsible for the Sun King's extensive building campaign, in Paris and Versailles particularly, may explain why he devoted his time to astronomy and the production of his *Cartes du Ciel* (*Charts of the Heavens*). In any event, his work is noteworthy for a number of reasons.

Royer includes on his map (plate 66) several new constellations. The figure of the French scepter and hand of justice, "Sceptrum," is placed at the feet of Pegasus and in such a position as to shine directly over the French capital when on the meridian of Paris. The symbol seems to have been created by Royer himself. Another symbol of French authority in the heavens is the constellation "Lilium" in the southern hemisphere (not shown here) representing the French *fleurs-de-lys* and replacing Keyzer's Musca (bee). The Lilium had first appeared in 1674. Also represented are the constellations of Tygris fluvius, Jordanus fluvius, and the giraffe (Camelopardalis) created by the Dutch cartographer Petrus Plancius. Several novas—new stars—are also identified, including Nova 1572 in Cassiopeia and Nova 1670 in Cyginus.

Royer's map reveals the architect's skill as a draftsman; the whole has a refined and elegant treatment of line in the rendering of Ursa major's fur and the folds of Andromeda's robes (if not in her hefty upraised arm). The faces of the figures are also worth noting, significantly more human in their expression than earlier depictions, i.e., the rather mournful face of Hercules and the stern and determined countenance of the peasant Boötes. These reflect current formulas of decorum adhered to in French academic art.

PLATE 66
Northern hemisphere
Augustin Royer (Paris, 1679)
Engraving on paper, diameter: 13½″

A number of choices were open to the celestial map-maker. The stars could be shown as seen from earth or from beyond the crystalline sphere. The heavens could be presented on a globe or on a flat surface. The map could be defined in terrestrial or celestial terms (either centered on the earth's poles or on the celestial poles). A major element affecting how the stars were to be projected was the zodiac. If the design was centered on the celestial poles to the ecliptic (as in Dürer's work) the zodiacal figures, theoretically, would be split in half, straddling as they do the line of the ecliptic. Dürer solved the problem by placing the zodiac figures in one hemisphere, spilling over the ecliptic boundary. On the other hand, the maps shown here are oriented so that the earth's equator is used as the boundary line encircling the spheres. This projection allows the ecliptic (the zodiac) to describe an arc through a portion of the circle of the projection, so all the zodiacal figures can be shown on a single plate.

Weighing all the advantages and disadvantages of the various methods, the choice of projection often seems to be, in the end, a matter of aesthetic taste. The maps shown here (plates 67, 68), produced about 1686 by the British mapmaker Philip Lea (fl. 1660–1700),

PLATE 67
Northern hemisphere
Philip Lea (London, c. 1686)
Engraving on paper, 14½ x 14½"

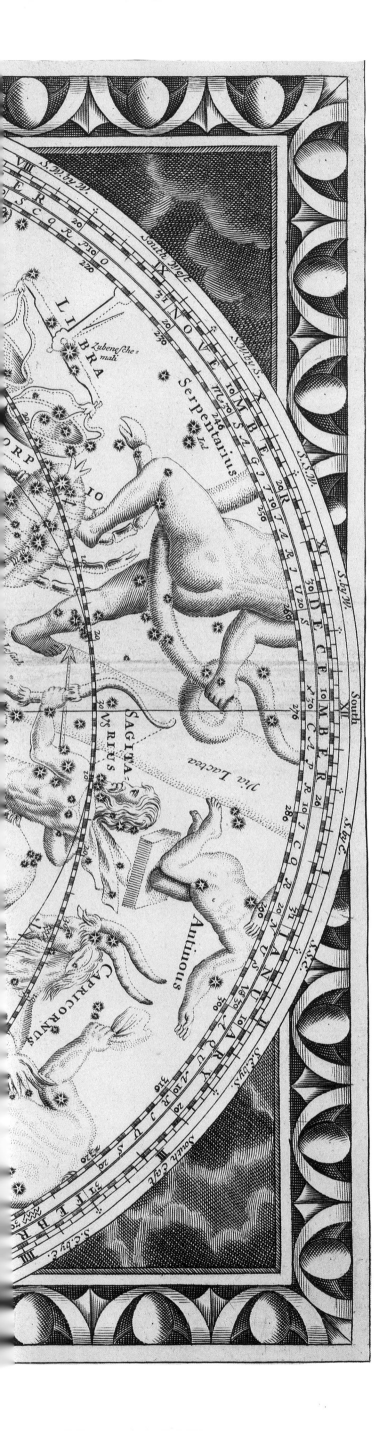

PLATE 68
Southern hemisphere
Philip Lea (London, c. 1686)
Engraving on paper, 14½ x 14½"

use a north- and south pole to equator projection. The resulting projection is more successful artistically in one hemisphere than in the other. The northern hemisphere is a rather attractive composition, especially in the configuration of Pegasus, Ursa major, and Leo, whose large, adequately rendered figures balance one another. The southern hemisphere is less successful. Virgo is shown sliced in half, and the cropped figures of Aquarius, Antinous, and Serpentarius, the snake-handler, are especially unfortunate, although the particular pose of the latter with his snake cannot be blamed entirely on Lea.

Lea's map depicts a constellation in the northern hemisphere worthy of note. At one time it was said to have been created by Halley in honor of Charles II, but is now known to have been put in the sky for another king and from another source. The constellation is "Cor Caroli Regis Martyris," honoring Charles I, who was executed on January 30, 1649 (on the Julian calendar). It is shown as a single star enclosed within a heart surmounted by a crown, in the region between Coma Berenices and Ursa major. This was neither the first nor the last time a king would be honored by a place in the heavens.

The decline of Italian supremacy in cartographic production, resulting from the shift in trade routes from the Mediterranean to the Atlantic, was matched in the seventeenth century by the rise of the Dutch mapmakers Ortelius, Mercator, Hondius, and the Blaeu family. Italy could nonetheless still claim Vincenzo Maria Coronelli (1650–1718), Cosmographer of the Venetian Republic, as one of the truly outstanding cartographers of the period.

Coronelli's terrestrial and celestial globes were known throughout Europe and admired for their elaborate beauty. His first pair of globes were produced in 1678 for the duke of Parma, and they so favorably impressed the French ambassador that Coronelli received an invitation to visit Paris, where he spent two years constructing a pair of globes for Louis XIV. These

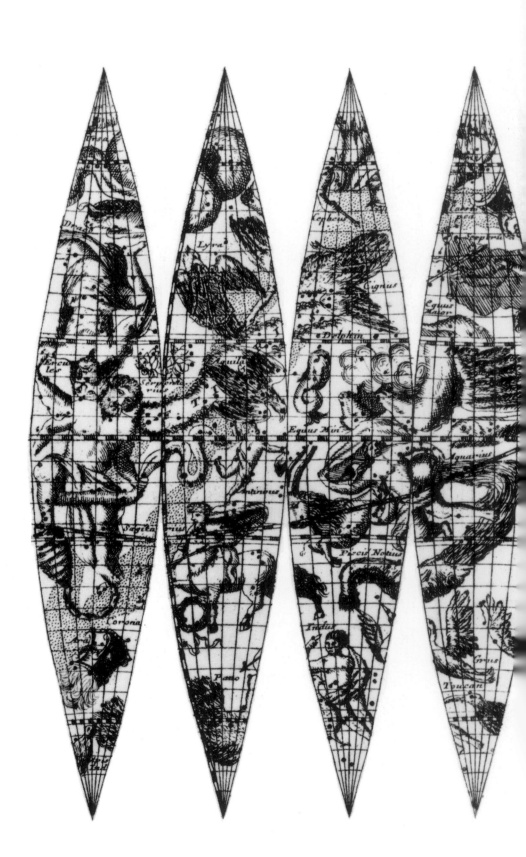

PLATE 69
Celestial globe gores
Vincenzo Maria Coronelli (Cologne, 1693)
Engraving on paper, 4 x 10″
From *Epitome Cosmografica*

enormous and ornate productions (fifteen feet in diameter) were completed in 1683 and remained the largest globes ever created until the 1920s. Now at Versailles, they were displayed by the French kings in various royal residences and the Royal National Library, appropriately grand and powerful images of divine right.

Founder of the geographical society Accademia Cosmografica degli Argonauti in Venice, Coronelli's work was not confined only to large-scale productions. He also created miniature globes three inches in diameter, as well as atlases and single maps. The set of gores presented here (plate 69) reveals the baroque style of his creations. Coronelli's work is significant for the quality and excellence of the presentation rather than the accuracy of its information.

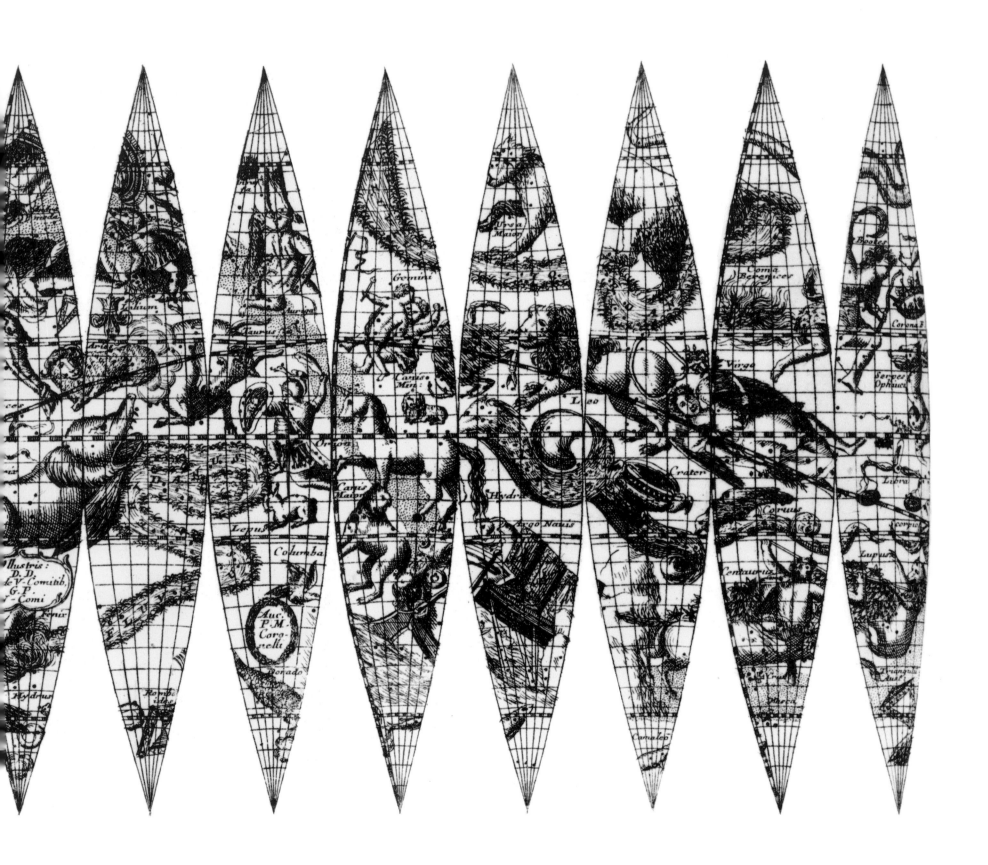

NEW HORIZONS

PLATE 70
Southern celestial hemisphere
James Barlow (London, c. 1790)
Engraving on paper, 14 x 14"

This late eighteenth-century map (plate 70), attributed to the British cartographer James Barlow (fl. c. 1780), indicates the extent to which observations of the southern hemisphere had expanded the Ptolemaic catalog during the age of exploration.

For ancient and medieval geographers and astronomers, the known world was a small part of what we know today. The areas below the equator, and certainly below the tropic of Capricorn, could only be imagined by scientists and mapmakers. It was conjectured that a great land mass existed in the south, if only to balance out aesthetically the arrangement of the earth's known land areas. But the discovery and charting of the continent of Australia occupies a point in history far removed from Ptolemy. Australia's first appearance on terrestrial maps actually came after the southern stars had been charted. One could see farther in the sky than one could travel by sea, though there were limits to both kinds of exploration.

Still, the world at the time this map was made, about 1790, was fast becoming known. When Tasmania became a British possession in 1803, the stars overhead had already been charted and were waiting to greet settlers of the expanding British empire. Among the new constellations were those recorded by the Dutch navigator Keyzer, which so well reflected the fauna and other phenomena of the New World—constellations called Musca (bee), Avis Indica (bird of paradise), Chameleon, Dorado (swordfish), Grus (crane), Hydrus (watersnake), Indus (Indian), Pavo (peacock), Phoenix, Piscis volans (flying fish), Toucan, and Triangulum Australe. Other constellations of the seventeenth century depicted here are the Columba Noachi of Petrus Plancius (Noah's dove); the Robur Carolinum, named by Halley and symbolizing the royal oak at Boscobel which hid Charles II from the Republican forces; and the Crux, here "Crosiers" or Southern Cross, caught in the rear legs of the Centaur, conjured by Portuguese and Spanish sailors, near the southern pole. Strange and wonderful beasts and forms were seen in the southern skies substituting for the mythological images explorers knew in their northern birthplaces.

Later, more constellations would be added, cluttering the sky with even stranger and more modern appurtenances. The French cartographer Nicolas Louis de Lacaille (1713–1762) had created a number of new representations for the southern hemisphere already in the *Coelum australe stelliferum* (*Charts of the Stars and the Heavens*) (published posthumously in 1763), including such appropriate but unlikely mechanisms as Fornax chemica (chemical furnace), Horologium (clock), Pyxis nautica (mariner's compass), Antlia pneumatica (air pump), Telescopium (telescope), and Microscopium (microscope). Some of these new images— symbols of a scientific, industrial age—would disappear; a few would remain.

Not all the stars that formed new constellations, it should be noted, were discovered since the age of the telescope. Ptolemy lists a number of stars that did not fit into his forty-eight constellations, including them as "unformed stars." As they continued to shine in the sky, other astronomers such as Barlow placed them in new constellations made possible by telescopic observation and southern exploration.

The success of the Cellarius atlas was due to its comprehensive coverage of all available and divergent aspects of astronomy combined with the highest artistic achievement. The geocentric and heliocentric arrangements of the universe were presented along with Tycho Brahe's compromise system; the celestial depictions of Schiller's Christianized heaven were offered along with ancient and contemporary views of the constellations. More than just a star atlas, the work of Cellarius was a monument to the advancement of astronomical science from ancient times to the seventeenth century, and beyond. The *Atlas Coelestis* presented not only star positions, it set forth the universal vision of Western thought in all facets.

A testimony to the greatness and popularity of the Cellarius atlas is that in the middle of the eighteenth century, almost one hundred years after it first appeared, the Amsterdam map publishing house of Gerard Valck and Petrus Schenck would republish the work. Enough astronomical observations and discoveries had been made in the interim, however, to make the reproduction obsolete. Moreover, plate 71 reveals that the new work was not entirely faithful to the original, substituting pastel pinks and greens and pale golden yellows for the brilliant shades preferred during the eighteenth century.

Given the long-lived belief in the geocentric universe, which continued to be held well after Copernicus had posited a heliocentric system, it is not surprising that the Cellarius atlas would be reissued without updating. The lag between popular acceptance and scientific discovery was—and in some respects still is—immense. The significance of this lag, however, in considering the history of celestial cartography, cannot be over-emphasized. Throughout its development, the history of maps of the heavens reflects a slow awakening of mankind's consciousness in the wake of scientific advancement.

Astronomy had paved the way for the development of modern science in the sixteenth and seventeenth centuries. Astronomical discoveries and improvements in scientific instruments such as the telescope laid the foundation for later theoretical and practical applications. The Valck and Schenck reprint of Cellarius, however, demonstrates that while science expanded the horizons of the universe, older conceptions continued to maintain their hold on the popular imagination.

PLATE 71
Northern celestial hemisphere
Andreas Cellarius (Amsterdam, c. 1705)
Engraving on paper, hand colored, 18 x 22″
From Gerard Valck, Peter Schenck's reprint of
Atlas Coelestis seu Harmonia Macrocosmica

By the time this globe (plate 72) was created by John and William Cary in London in 1800, the age of celestial cartography as presented in these pages was drawing to a close. Celestial maps and globes would, of course, continue to be produced, but scientific advancement in the nineteenth century would bring forth so vast a quantity of information to be depicted on a map's surface that other forms of projection would be developed to convey the data. The heavens in 1800 were already quite crowded, and the fantasy of star travel and exploration was turning into fiction what in our own century would become reality. But plenty of people still looked to the sky for constellations, seeking out the images depicted on the Cary globe, as indeed we continue to do.

The delicate balance, even in 1800, between the artist and the scientist, the astronomer and the church, between theory and data, knowledge and faith, was beginning to waver. In a few decades, Darwin would change our views of the natural world, Jules Verne would have us rethink the possibilities of cosmic frontiers, and Nietzche would have us look at the night sky as an empty abyss which ultimately would look back at us in its emptiness.

The Cary globe, however, still belongs to a comforting tradition of celestial cartography many hundreds of years old. "Cary's New Celestial Globe on which are correctly laid down upwards of 3500 stars," reads the globe's cartouche, "selected from the most accurate observations and calculated for the Year 1800 with the extent of each constellation defined by Mr. Gilpin of the Royal Society." The globe's makers, William and John Cary, were well suited for the task. William was a member of the Astronomical Society and well known for the instruments he made—including sextants, microscopes, and reflecting and refracting telescopes. John was a highly respected cartographer. Their work was widely admired for its accuracy and beauty.

The Cary globe, in a clean and attractive fashion, depicts the new and old constellations. In the detail (plate 73) can be seen, for example, the constellation "Taurus Poniatowskii," introduced in 1777 by Martin Poczobut of Vilna in honor of Stanislas II, King of Poland (seen just below and right of the cartouche). This constellation is only one of a number which were swept away in later depictions as the proliferation of new constellations waned and a modern approach to the sky and galaxies and the universe evolved in the nineteenth century.

PLATE 72
"Cary's New Celestial Globe"
John Cary, William Cary (London, 1800)
Engraving on paper, hand colored, formed into a globe, on stand, diameter: 12″

PLATE 73
"Cary's New Celestial Globe," detail

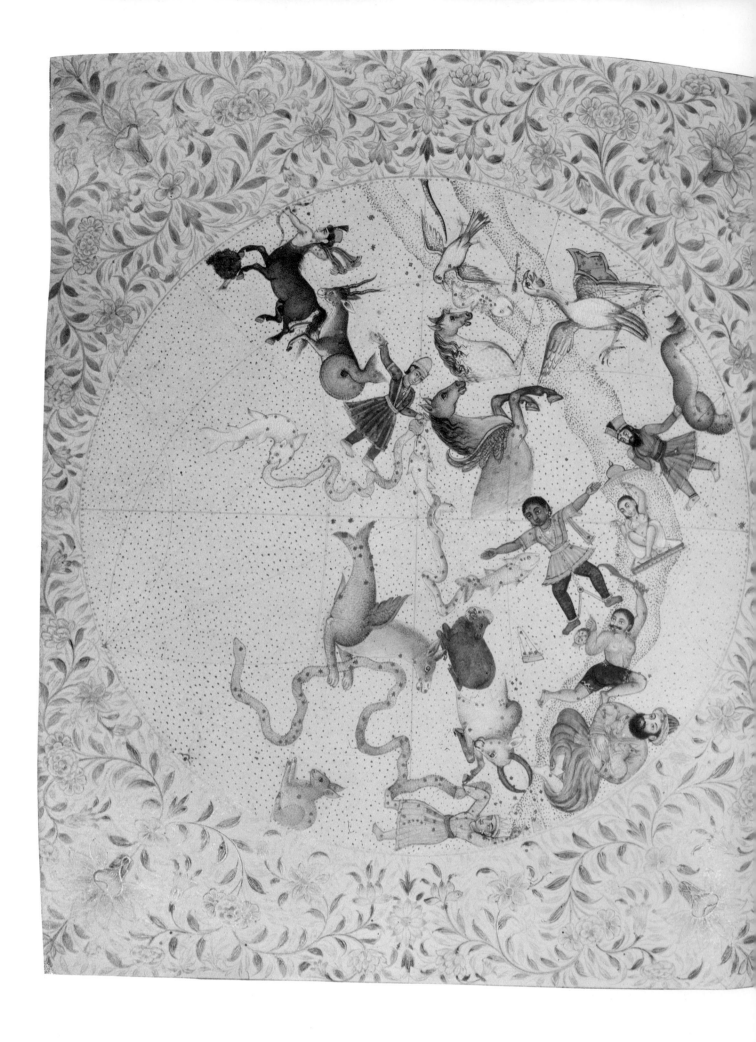

The history of astronomy and the production of celestial charts, as has been seen, is a history of migration and translation—from the Babylonians to the Greeks to the Islamic world to the medieval astronomers of Europe to modern times. The development of a historical theme or subject is never in a straight line, and nowhere, perhaps, is this more true than in India, where, despite continual exposure to outside influences, an approach to the charting of the heavens evolved that is unlike any other in the world.

In this delightful celestial composition (plate 74),

produced in 1840 at the time of English dominance in India, the familiar constellations are presented in the most unfamiliar costumes. In light of the Indian approach to science as a whole, the representation here is rather an interesting comment on the demise in the West of astrology, and on the impending demise of celestial cartography in the modern world. Produced forty years after the Cary globe, this work, in Sanskrit, by Durgasankara Pathaka from Benares, was ordered by Khadga Singh as part of a manuscript horoscope for his son, Prince Navanihal Singh. If in Europe rational-

ism was soon to wipe out any trace of superstition bound up in the study or practice of astrology, this treatise on the zodiac might have given the remaining Western believers new faith. The entire work is entitled *Sarvas-iddhantatattvacudamani*, or *Jewel of the Essence of All Sciences*. Including all the major constellations which could be seen from India, the work could well stand as a reminder that the interpretation of the sky in symbolic terms extends far beyond bounds of logic or knowledge about the world as we know it, or believe that we know it.

PLATE 74
Sanskrit star map
Durgasankara Pathaka (India, 1840)
Manuscript on paper, 16 x 20"

PLATE 75
System of the universe
Thomas Wright (London, 1750)
Engraving on paper, 9 x 7″
From *An Original Theory or New Hypothesis of the Universe*

As the telescope expanded mankind's vision of the sky overhead, astronomers continued to expand the limits of the tools available to them. Increasingly sophisticated instruments led to more refined images of the heavens and speculation about what might lay beyond the range of telescopes. Most eighteenth-century scientists and astronomers were theists, and their inquiries and theorizing were often understood to be exercises in proving the presence of God in His creation. Isaac Newton's mysticism was a powerful influence on scientists of the time, the vast majority of whom held to the belief in "Natural Theology"—a belief that the existence of God could be demonstrated from the study of the natural world. The increase in scientific knowledge meant, therefore, a reinforcement of belief in religious conviction.

One theist-scientist was Thomas Wright (1711–1786), an astronomer from Durham whose famous work, *An Original Theory or New Hypothesis of the Universe*, though published in the mid-eighteenth century (London, 1750), is in many respects a final, brilliant effort to incorporate astronomical discoveries and observations into a positive proof of God's existence. At the same time, Wright attempted to step beyond the limits imposed by the scientific knowledge of the time, producing an image of the universe both distant from and related to the one presented earlier in plate 2.

In his introduction to *Original Theory*, Wright states that knowledge of God is inadequate without knowledge of astronomy, returning to one of the basic motivations for creating star charts in the first place: that belief complemented by observation would produce a coherent vision of the heavens. Once belief in Natural Theology declined and scientific data was assured an even more important role, that vision would change once again.

Often credited with the first accurate explanation of the Milky Way, Wright described it as the arc of the disk-shape galaxy in which our solar system is located. In the plane of the disk (plate 75) the stars appear as a band of light. This interpretation of Wright's work, however, must be credited to Immanuel Kant, for Wright's main objective was to venture beyond astronomical explanations. It was Kant, in fact, who interpolated Wright's work as a description of the Milky Way.

For Wright, the divine center of the universe was the moral and gravitational center. Moreover, he posited a multiplicity of systems within the universe, a collection of "star systems," as shown here, each with its individual divine center. According to Wright, one could hope to be reincarnated in one of these systems, closer to the ultimate divine center. Wright eventually revised his *Original Theory*, attempting to further reconcile his belief in an afterlife with a universal design, but the revision remained unpublished at his death. Ironically the beautiful plates in his *Original Theory* and his other work, *Clavis Coelestis* (*Key to the Heavens*), showed that he was a far greater artist than either scientist or theologian, and they were probably more effective than his ideas in attracting his original subscribers, as well as his modern collectors.

LIST OF PLATES

FURTHER READING

Bagrow, Leo. *History of Cartography*. Revised and enlarged by R. A. Skelton. Cambridge: Harvard University Press, 1966.

Brown, Lloyd A. *The Story of Maps*. Boston: Little, Brown and Co., 1949.

D. S. B. *Dictionary of Scientific Biography*, 16 volumes. New York: Charles Scribner's Sons, 1970–80.

Hamilton, Edith. *Mythology*. New York: Grosset Universal Library, 1942.

Horblit, Harrison D. *One Hundred Books Famous in Science*. New York: The Grolier Club, 1964.

Houzeau, J. C., and Lancaster, A. *Bibliographie generale de l'astronomie ou catalogue méthodique des bouvrages, de mémoires et des observations astronomiques publiés depuis l'origine de l'imprimerie 1880*, 3 volumes. Revised edition by D. W. Dewhirst. London: Holland Press, 1964.

Klebs, Arnold C. *Incunabula Scientifica et Medica*, volume 4, part 1. Bruges: Osiris, 1938.

Knight, David M. *Natural Science Books in English, 1600–1900*. New York: Praeger, 1972.

Koeman, C. *Atlantes Neelandici*. Amsterdam: Theatrum Orbis Terrarum, Ltd., 1967–71.

Poggendorff, Johann C. *Biographisch-Literarisches Handworterbuch zur Geschicte der Exacten Wissenshaften*, 2 volumes. Leipzig: Verlag von J. A. Barth, 1863. Reprinted, New York: Walter Johnson, 1974.

Riccardi, Pietro. *Bibliotheca Matematica Italiana*. Modena: Societa Typografica, 1878–93.

Sarton, George. *Introduction to the History of Science*. Baltimore: Williams and Wilkin for Carnegie Institution, 1927–48.

Stillwell, Margaret B. *The Awakening Interest in Science during the First Century of Printing*. New York: Bibliographical Society of America, 1970.

Thorndike, Lyn. *A History of Magic and Experimental Science*, 8 volumes. New York: Columbia University Press, 1923–58.

Warner, Deborah J. *The Sky Explored*. New York: Alan R. Liss, Inc., 1979.

Wolf, Abraham. *A History of Science and Technology*, 2 volumes. New York: Harper Torchbooks, 1950–52.

Zinner, Ernst. *Deutsche und Niederlandische Astronomische Instrumente des 11–18 Jahrhunderts*. Munich: C. H. Bech'sche Verlagsbuchhandlung, 1956.

ACKNOWLEDGMENTS

The author wishes to express his thanks to the museums, libraries, and private collectors who granted permission to reproduce from their collections the images that appear in this book:

Beineke Rare Book and Manuscript Library, Yale University: plate 75; Biblioteca Apostilíca Vaticana, Rome: plate 21; The British Library, London: plates 7, 8, 15, 23, 24, 29, 30, 32, 42, 43, 44, 48, 49, 50, 54, 55, 62, 63, 64, 65, 66, 67, 68, 70, 74; The Houghton Library, Harvard University: plate 61; Musée Condé, Chantilly, Giraudon/Art Resource, New York: plate 3; Nederlandsch Historisch Scheepvart Museum, Amsterdam: plates 59, 60; The New York Public Library, Rare Book Room: plates 33, 34, 35, 37, 38, 39, 40, 41, 69; The New York Public Library, Science and Technology Research Center, Astor, Lenox and Tilden Foundation: plate 47; The New York Public Library, Spencer Collection: plates 4, 9, 10, 11, 12, 13, 14, 16, 51, 52, 53, 56, 57, 58; The Pierpont Morgan Library, New York: plates 5, 6, 17, 18, 22, 25, 26, 27, 28; The Pierpont Morgan Library, New York, Courtesy George Braziller, Inc.: plate 1; Private Collection, Courtesy Sotheby Parke Bernet: frontispiece, plates 2, 20, 31, 36, 45, 46, 71, 72, 73; Rare Book and Manuscript Library, Columbia University: plate 19